Heartland Tobacco War

Heartland Tobacco War

Michael S. Givel and Andrew L. Spivak

Forword by Stanton A. Glantz

LEXINGTON BOOKS
Lanham • Boulder • New York • Toronto • Plymouth, UK

Published by Lexington Books
A wholly owned subsidiary of The Rowman & Littlefield Publishing Group, Inc.
4501 Forbes Boulevard, Suite 200, Lanham, Maryland 20706
www.rowman.com

10 Thornbury Road, Plymouth PL6 7PP, United Kingdom

British Library Cataloguing in Publication Information Available

Library of Congress Cataloging-in-Publication Data

Givel, Michael, 1954–
Heartland tobacco war / Michael S. Givel and Andrew L. Spivak.
p. ; cm.
Includes bibliographical references and index.
ISBN 978-0-7391-7692-4 (cloth : alk. paper) — ISBN 978-0-7391-7693-1 (electronic)
I. Spivak, Andrew Lawrence, 1972– II. Title.
[DNLM: 1. Tobacco Smoke Pollution—history—Oklahoma. 2. Tobacco Smoke Pollution—legislation & jurisprudence—Oklahoma. 3. Government Regulation—history—Oklahoma. 4. History, 20th Century—Oklahoma. 5. History, 21st Century—Oklahoma. 6. Tobacco Industry—legislation & jurisprudence—Oklahoma. WA 33 GI6]
RA1242.T6
362.29'609766—dc23 2013016663

∞™ The paper used in this publication meets the minimum requirements of American National Standard for Information Sciences—Permanence of Paper for Printed Library Materials, ANSI/NISO Z39.48-1992.

Printed in the United States of America

Contents

List of Figures and Tables vii

Foreword ix

Acknowledgments xi

1 Introduction: A Tobacco War in America's Heartland 1

2 Historical Background of Tobacco Control in Oklahoma 9

3 The Influence of the Tobacco Industry Lobby 19

4 Leslie Beitsch and the 2001–2003 Clean Indoor Air Campaign 29

5 The 2004 Battle for Cigarette Taxes 55

6 Aftermath: Clean Indoor Air Post-2003 73

7 Lessons Learned: Public Ethics and the Public Good 93

References 101

Index 113

About the Authors 121

List of Figures and Tables

FIGURES

Figure 1.1	Per capita cigarette consumption in Oklahoma, compared with the United States and California.	3
Figure 1.2	Smoking behavior responses of 647 randomly surveyed Oklahomans May 10–28, 2004 (63.2% of respondents male, 33.4% female, 3.4% unknown).	4
Figure 4.1	David Simpson—12/18/2002. *Tulsa World.* Reprinted with permission (not an endorsement).	49
Figure 5.1	Bar chart of responses to May 2004 random sample Oklahoma statewide survey.	67
Figure 5.2	Advertisement from the Vote No on Question 713 campaign.	70
Figure 5.3	Advertisement from the Vote Yes on State Question 713 campaign.	71
Figure 6.1	2008: Opinion on allowing smoking inside bars, by frequency respondents visited bars.	76
Figure 6.2	Opinions about smoking inside restaurants, shopping malls, and bars.	77
Figure 6.3	Opinions on smoking in indoor work areas from 2002 to 2008.	78
Figure 6.4	Employers' official smoking policy for work areas from 2002 to 2008.	79
Figure 6.5	Contributions to political campaigns, political parties, sponsorship of balls, and other PACs by the tobacco industry PACs' direct contributions per annum.	85

TABLES

Table 3.1	Lobbyist or Other Person Gift Report (January 17, 2001–January 15, 2003).	22
Table 3.2	Lobbyist or Other Person Gift Report, Attachment (July 18, 2002).	24
Table 3.3	Contributions to Leadership of the Forty-Ninth (2003–2004) Oklahoma Legislature, 1995–2004.	26
Table 5.1	Oklahoma Ethics Commission, Form C-1R (November 10, 2004).	69
Table 6.1	Bivariate Pearson Correlations.	90
Table 6.2	Stepwise OLS Regressions: Dependent Variable: Tobacco Policy Score. Unstandardized (B) and Standardized (STB) Coefficients	91

Foreword

On the surface Oklahoma and California couldn't seem more different: Oklahoma is a small staunchly red state in America's heartland and California is, well, California. But Givel and Spivak show that on one dimension—tobacco control—they are two peas in a pod. Even a cursory comparison of *Heartland Tobacco War* with our 2000 book *Tobacco War: Inside the California Battles* (University of California Press) shows that the lessons are the same.

The tobacco companies are well-practiced at hiring well-connected lobbyists, buying off key legislative leadership and hiding behind "third party" allies (most notably the Oklahoma Restaurant Association) to stay out of the public eye, and manipulating the process to kill effective tobacco control policies despite strong public support for these measures. As long as the health forces took cautious insider approaches (most notably quiet, behind-the-scenes lobbying) they lost because they were playing on the industry's turf.

On the other hand, when individuals and organizations—I'm leaving the details out so to not spoil the story—were willing to take personal and organizational risks by moving outside the confines of the legislature, take on Big Tobacco's third party allies, mobilize public support, and recognize that fighting Big Tobacco almost always includes a detour through the courts, they often won.

Heartland Tobacco War has one big advantage over our earlier book on California: Full access to millions of pages of previously secret tobacco industry documents. These documents illuminate just how aggressive and nasty the cigarette companies are when dealing with their political opponents. This information, combined with first person interviews with key participants, allow Givel and Spivak to include several specific examples of how the tobacco companies use campaign contributions to reward and punish friends and opponents, as well as the bare knuckles threats of budgetary retaliation when uppity public health officials take their responsibility for leading efforts that will actually reduce smoking. These are remarkable tales that anyone willing to fight tobacco needs to read and understand that these obstacles can be overcome.

Tobacco control is not for the faint of heart. *Heartland Tobacco War* is one more illustration of how with courage, intelligence, persistency, and a little luck, public health can beat Big Tobacco.

Stanton A. Glantz, PhD
Professor of Medicine
Director, Center for Tobacco Control Research and Education
University of California San Francisco

May 2013

Acknowledgments

First, we would like to recognize American Heart Association Grant #0335002N, which facilitated our original research between 2003 and 2005 on tobacco policy in Oklahoma and was the first step that eventually led to writing this book. Additionally, the Oklahoma Tobacco Research Center Grant RS20091891-07 between 2010 and 2011 assisted us in the research and writing of chapter 6, and the University of Oklahoma's Center for Research Program Development and Enrichment, Faculty Investment Program was likewise an important source of support for this project.

We also want to acknowledge the tireless research and editorial assistance of University of Oklahoma doctoral students Ami Stearns and Ivy Shen.

Finally, a number of colleagues and professionals volunteered their time to be interviewed, and each of them provided valuable insight: Bob Minor, Doug Matheny, Drew Edmondson, Laura Beebe, Richard Barnes, Joy Leuthard, Alice Walker, and Barbara Kumpke. We are especially grateful for the extended time that Leslie Beitsch gave us across two interviews, and we hope that this book adequately reflects his extraordinary work as health commissioner.

Chapter One

Introduction

A Tobacco War in America's Heartland

Through the end of the twentieth century, tobacco industry insiders and lobbyists carried on a close, cooperative affair with the state of Oklahoma, going so far as to describe the relationship between the Industry and its Oklahoma allies as a "love fest" (Philip Morris 1989). Compared to many other states, virtually no progress in tobacco control regulations was made during the 1980s and 1990s (Spivak and Givel 2005; Givel and Spivak 2008). All of that was about to change with the 2001 arrival of a new state health commissioner, Dr. Leslie Beitsch, who publicly declared his intention for an all-out attack on tobacco through vigorous advocacy for new legislation to promote clean indoor air and cigarette tax increases.

The period during which Oklahoma's controversial, maverick health commissioner waged this battle against tobacco interests was a critical turning point for a state that had spent years behind the nation on health policy. Not only was he victorious in the arena of clean indoor air, he also set the stage for increasing cigarette taxes to levels more on par with other states. This book is intended not only as a chronicle of the political and public policy drama that surrounded Dr. Beitsch's years in office, but also an examination of the way state health policy change took place. We discuss themes including how and why the health commissioner's aggressive public and political advocacy for rigorous clean indoor air legislation in 2001 to 2002—independent of the legislature and governor—also reflect the promotion of public administration ethics (e.g., compassion, benevolence, fairness, and optimism). In the context of public administration theory, we also analyze the health commissioner's campaign, which held politicians publicly accountable for their actions and promoted the public interest. Finally, we review the ways that bypassing the state legislature with outside pressure tactics in a highly aggressive *outsider* public campaign resulted in realized success; in 2003, substantial clean indoor legislation was finally passed in a very conservative mid-south state.

Oklahoma's tobacco war also has a lot to say, in our opinions, about political theory. Using a framework of bureaucratic activism, utilitarianism and the Public Administration Ethics Code, as well as the politics of tobacco and the tactics of the tobacco industry, we argue that Dr. Beitsch's aggressive public campaign promoted anti-tobacco policy by defining public interest as the desire for well-being and happiness for the greatest number of people. His advocacy encompassed the Code's tenets of compassion, benevolence, fairness, and optimism (Givel and Spivak Winter 2011–2012).

OVERVIEW

In 1985, the R. J. Reynolds Tobacco Company confidently announced in an internal political analysis of Oklahoma that "no piece of smoking restriction legislation has ever been voted on by either house of the Oklahoma State Legislature," and that "groups interested in pursuing passage of smoking restrictions are either assuming a low profile or are virtually nonexistent" (R. J. Reynolds 1985). Over the next two decades, the tobacco industry would play a key political role in fighting public health initiatives to regulate tobacco in Oklahoma. Thus, efforts to enact smoking restrictions in public and work places were slow and unsuccessful, even resulting in a 1987 act that actually required smoking sections in restaurants. Additionally, under the influence of Tobacco Institute lobbying in 1986, the state passed legislation for the "preemption" of local tobacco control laws, which prevents local municipalities from enacting stronger clean indoor air laws and youth access prevention measures, thus effectively routing all such efforts through the state legislature, where the industry had firm control (Tobacco Institute 1994). This preemption law remains in effect today.

Reflecting the dismal state of tobacco legislation in the state, the Oklahoma Tobacco Use Prevention and Cessation Advisory Committee, an advocacy group with contributors from the state Department of Health, other state agencies, and health groups such as the American Heart and Lung Associations, published a report in May 2002 titled *The Oklahoma State Plan for Tobacco Use Prevention and Cessation*. The Plan's comprehensive policy proposals included restrictions on tobacco sales in vending machines, stiffer penalties for tobacco sales to minors, repeal of state preemption laws, adopting clean indoor air laws in workplaces, increasing state tobacco excise taxes, conducting vigorous anti-tobacco counter-marketing efforts, cessation programs, and community education efforts (Oklahoma Tobacco Use Prevention and Cessation Advisory Committee 2002). All these recommendations were exemplary reflections of a health agenda that most public health scholars and tobacco control advocates had been promoting for years, especially clean indoor air and cigarette tax increases, but thus far actual progress toward these goals had been nonexistent.

Not surprisingly, the State of Oklahoma's record of health and tobacco use into the early 2000s coincides with the cozy status that the tobacco industry enjoyed there. Oklahoma's 2002 per capita sales of cigarettes were 27 percent higher than the national average, and 183 percent higher than California (figure 1.1). In contrast, California was able to enforce a rigorous tobacco control program between 1989 and 1993, which led to a rapid decline in smoking prevalence (Glantz and Balbach 2000). Even when the large and aggressive program was reduced in funding during 1994–1998, due to the industry's success cultivating allies such as the California Medical Association, per capita cigarette consumption continued to decline (Glantz and Balbach 2000). Oklahoma's lack of stronger tobacco control measures, especially when compared with California, certainly contributed to the large difference in figure 1.1 (Bitton, Fichtenberg, and Glantz 2001).

In the early 2000s, Oklahomans identified themselves as smokers at slightly higher rates than the rest of the nation. The Oklahoma State Department of Health's 2004 *State of the State's Health* indicated that while the proportion of cigarette smokers among the national adult populace was about 23 percent in 2002, almost 27 percent of Oklahoma adults indicated current cigarette use (Oklahoma State Department of Health 2004). A May 2004

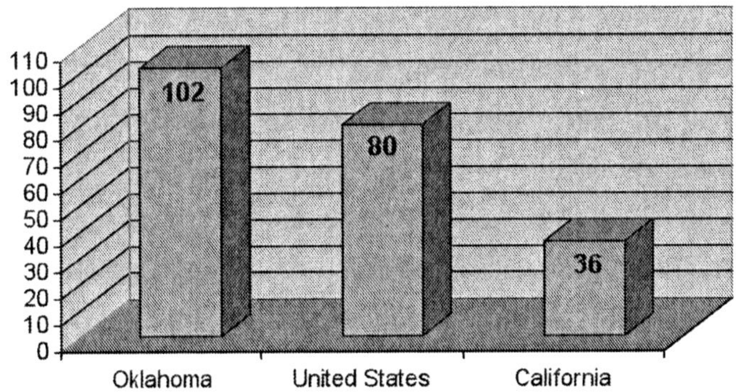

Figure 1.1. Per capita cigarette consumption in Oklahoma, compared with the United States and California.
Source: Oklahoma State Department of Health, Tobacco Use Prevention Service, 2004. Retrieved from http://www.health.state.ok.us/program/tobac/factsand.htm

statewide telephone poll conducted by the Oklahoma Public Opinion Learning Laboratory at the University of Oklahoma (OU POLL) confirmed the rate for Oklahomans, finding 26.3 percent smokers among the 647 interviews (Outwater 2004).

Not only were more than one-fourth of Oklahomans current smokers, but an additional 27.4 percent of respondents to the OU POLL survey (figure 1.2) identified themselves as former smokers, and men were much more likely to be both current and former smokers than women. Additionally, 51 percent of Oklahoma smokers stated that they had begun by the age of sixteen (Outwater 2004). By 2006, this trend had remained steady with 25.1 percent of Oklahoma adults currently smoking cigarettes at least once every month (Oklahoma State Department of Health 2012).

By 2007, the health impact of smoking in Oklahoma would become staggering; 5,800 adults in Oklahoma died from tobacco-related illness annually, one of the largest percentages of smoking related deaths in the United States (Oklahoma State Department of Health 2007). Cigarette sales represented 87 packs per person each year in Oklahoma compared to 71 packs per per-

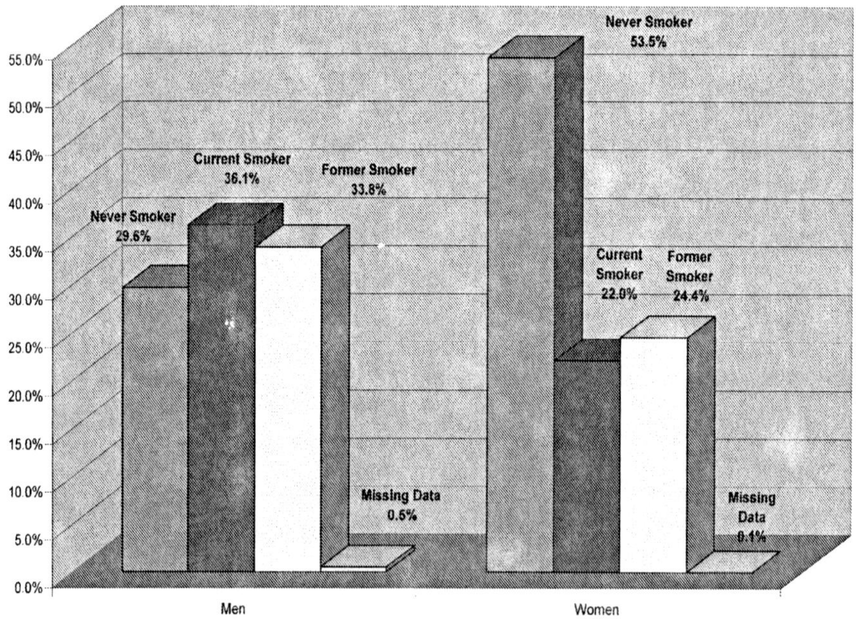

Figure 1.2. Smoking behavior responses of 647 randomly surveyed Oklahomans May 10–28, 2004 (63.2% of respondents male, 33.4% female, 3.4% unknown).
Source: Tobacco Use and Legislation Survey (May 10–28, 2004). University of Oklahoma Public Opinion Learning Laboratory (OU POLL).

son each year nationally, and 33 packs per person each year in California (Oklahoma State Department of Health 2007). The state health department estimated that 87,000 children currently under 18 years old will die prematurely due to tobacco addiction (Oklahoma State Department of Health 2007). Among women who recently gave birth in Oklahoma, about one-third smoked prior to pregnancy, with 42 percent quitting smoking in their last trimester of pregnancy (Oklahoma State Department of Health 2007). The costs of health care for tobacco related illness is over $2 billion a year (Oklahoma State Department of Health 2007).

What accounted for the poor state of health and the favored position of tobacco interests in Oklahoma, particularly in the area of clean indoor air? One explanation is that in the early 2000s for each dollar spent on tobacco prevention and cessation in Oklahoma, the industry spent about $35 for lobbying to support policies favorable to the industry (Oklahoma Tobacco Settlement Endowment Trust Fund 2002). However, the industry's sophistication extends beyond lobbying expenditures; it seeks information to better understand Oklahoma government processes and health advocacy activities in order to refine its anti-tobacco control activities. In 1998, Philip Morris contracted a 57 page political consulting report titled "Oklahoma Tobacco Regulation and Control" from the firm Richardson Ziebart Consulting, LLC, which outlined government bodies in the state and how they operated, the progress of tobacco control, federal involvement in tobacco control, and the structure and activities of health groups (Richardson Ziebart Consulting LLC 1998).

LESLIE BEITSCH AND THE 2003 VICTORY

Real change in clean indoor air did not arrive until 2003, when Senate Joint Resolution 21—created in reaction to the new health commissioner's aggressive two-year public campaign—brought about smoke-free public places and workplaces (Tobacco Institute 1990a; Oklahoma State Senate 2003a). This legislation finally gave Oklahoma a prohibition on smoking in public enclosed areas that are public places, as well as all indoor workplaces, except designated "smoking rooms" in restaurants with separate ventilation and negative air draft. Restaurants finally complied by the March 2006 deadline (Oklahoma State Senate 2003a).

The road toward this achievement was not only grueling, but reflected unique challenges, as well as unorthodox strategies, in battling the tobacco industry through means not previously employed. The new health commissioner began by banning tobacco at all state health department buildings in January 2002. Then, in tandem with the Oklahoma Health Commission, he imposed a set of administrative rules that banned smoking in health care facilities and a second set of rules that covered most public places with some exceptions such as designated smoking areas. Most of the first and second set of rules were eventually approved by then-governor Frank Keating. Viewed as aggressive and unconstitutional boundary-breaking by the opposition, two businesses affiliated with the Oklahoma Restaurant Association (ORA), a tobacco industry lobby ally, filed lawsuits against the health department and the governor, claiming that the department's emergency ruling was part of an agenda to create a bureaucratic dictatorship (Talley 2002a, 2002c). The American Lung Association then filed a suit against the ORA plaintiffs, the ORA itself, and the health department, for allowing smoking in restaurants which, the suit alleged, violated the Americans with Disabilities Act (Harper 2002).

Both lawsuits died in the aftermath of Senate Joint Resolution 21, a tough new clean indoor air bill that ironically began with an endorsement by the ORA in a press conference a mere six months after their pro-tobacco suit against the health department (Oklahoma Restaurant Association 2002b). The dramatic legal and public relations struggles associated with Dr. Beitsch's campaign and this eventual triumph are the central foci of the story told in this book.

DATA AND METHODS

Our narrative will include extensive inside information on the Oklahoma tobacco battles in the early 2000s over clean indoor air restrictions and higher tobacco taxes. Data were obtained from primary sources including media (e.g., newspapers, journals, magazines) and government documents from the Oklahoma Tobacco Settlement Endowment Trust and Oklahoma State Department of Health obtained through Oklahoma Open Records requests. Additionally, we collected ethnographic data from in-depth interviews with key actors including health department officials, health department activists, state's attorney general Drew Edmonson, and former Health Commissioner Leslie Beitsch. These interviews include never before revealed insider facts from Dr. Beitsch, who resigned amidst the controversy and political pressure generated from the clean air movement. Also utilized are confidential tobacco industry documents located at the University of California, San Francisco Legacy Tobacco Document's Library online at http://legacy.library.ucsf.edu/ pertaining to industry lobbying strategy plans. Finally, we integrate several forms of descriptive and inferential data pertaining to health and tobacco use, public attitudes about tobacco control, tobacco lobby activity, and state legislator behavior.

All of these primary data sources are cross-checked or triangulated against each other to minimize bias and enhance validity. We attempt to counter potential bias error by analyzing qualitative and quantitative data through systematic and reproducible historical and archival methodology. This process also includes paying attention to obtaining data from primary sources and proper contextualization by incorporating the data into the narrative chain of events that were part of (1) clean indoor air policy from 2001 to 2003, (2) the subsequent tobacco tax increase in 2004, and (3) the progress of tobacco control in Oklahoma in more recent years. Additionally, all of our data will be made completely transparent for inspection, use, and replication by any interested party.

ORGANIZATION OF THE BOOK

The chapters in this book are organized in the following manner. After the introduction, the second chapter will cover the history of tobacco control activism, specifically the evolution of bureaucratic activism. The third chapter details the role of the tobacco lobby and its allies in countering clean indoor air legislation. The fourth chapter will chronicle the battle to pass stronger clean indoor air rules and legislation. Chapter 5 will analyze the way stronger clean indoor air legislation led to higher tobacco taxes in 2004, a struggle caught up in rhetoric concerning "sin taxes" and government intrusion. In the sixth chapter, we will apply the framework of public ethics to the health commissioner's campaign, and in chapter 7, we will analyze the lessons learned from Oklahoma's tobacco war, particularly as it applies to other states.

Chapter Two

Historical Background of Tobacco Control in Oklahoma

There are 86 known species of Nicotiana, or the tobacco plant, worldwide with most species containing nicotine including all varieties of commercial tobacco (Marks 2007; Wheeler 1935; Goodspeed 1954). Native tobacco species appear in North and South America, southwest Africa, some Pacific Islands, and in Australia (Horton 1981). In Australia, for example, this includes 26 native species (Marks 2007; Burbridge 1960; Wheeler 1935; Horton 1981; Goodspeed 1954). Tobacco species have a wide range of floral and vegetative shapes (Clarkson et al. 2004; Wheeler 1935; Burbridge 1960; Horton 1981; Goodspeed 1954). The plants can vary in height at maturity from a few centimeters to over four or more meters high (Clarkson et al. 2004; Horton 1981; Burbridge 1960; Goodspeed 1954).

Many scholars and historians have written that the tobacco plant was first cultivated in the New World in Central and South America as early as 6000 B.C. (Rajasevasakta and Aiyer 1966; World Health Organization 2009; Borio 2007). Indigenous people in Central and South America began smoking tobacco by 1 B.C. (Rajasevasakta and Aiyer 1966; Borio 2007; World Health Organization 2009). Until 1492 when Christopher Columbus encountered tobacco smoking and brought the substance to Europe for cultivation and consumption, tobacco consumption was unknown in the Old World (Rajasevasakta and Aiyer 1966; Borio 2007; World Health Organization 2009). The genus Nicotiana was named after French Diplomat Jean Nicot de Villman who introduced the Nicotiana rustica variety of tobacco from Portugal to France in 1560 (Rajasevasakta and Aiyer 1966; Borio 2007; World Health Organization 2009; Clarkson et al. 2004). According to many scholars, tobacco products were first brought to China around 1530 and India around 1600 (Rajasevasakta and Aiyer 1966; Borio 2007; World Health Organization 2009).

As for tobacco use in other parts of the world, Borio has concluded:

> Although small amounts of nicotine may be found in some Old World plants, including belladonna and Nicotiana africana, and nicotine metabolites have been found in human remains and pipes in the Near East and Africa, there is no indication of habitual tobacco use in the Ancient world, on any continent save the Americas. (Borio 2007)

ALTERNATIVE INFORMATION ON HOW TOBACCO SPREAD

Possible evidence that tobacco may have been used independently of New World tobacco plants was presented in a 1993 scientifically peer reviewed article by Parsche et al. in the prestigious medical journal, *The Lancet* (Parsche, Balabanova, and Pirsig 1993; Balabanova, Parsche, and Pirsig 1992). In this study the researchers tested for evidence of drug use in 72 Peruvian mummies dated 200–1500 A.D., 11 Egyptian mummies dated 1072 B.C.–395 A.D., skeletal material from two Sudanese corpses dated 5000–4000 B.C. and 400–1400 A.D., and ten south German corpses dated 2500 B.C. (Parsche et al. 1993; Balabanova et al. 1992). The results of the study indicated that for the Egyptian mummies, there was nicotine in the hair of four, in the soft material of four, and in the bones of 11 (Balabanova et al. 1992; Parsche et al. 1993). The conclusion of this study was published without any context as to how the nicotine may have gotten there.

Criticisms of the study were primarily centered on issues of the methodology used in the research, whether the mummies were real, contamination issues, and concerns about the evidence (Moore, 1993; Parsche et al. 1993; Balbanova, 1994). Parsche et al. (1993) vigorously responded to each of these criticisms. In particular, Dr. Balbanova, one of the researchers and authors of the study who was also a highly experienced forensic chemist, argued that the nicotine apparently was incorporated during postmortem embalming procedures (Parsche et al. 1993). The conclusions of this study, while vigorously debated and disputed, are possibly in contrast to the dominant historical account of how tobacco consumption spread from the New World to the Old World.

HISTORICAL CONTEXT OF TOBACCO CONTROL IN OKLAHOMA

Due to strong progressive sentiments, the territorial government of Oklahoma—the territory became a state in 1907—enacted a total cigarette prohibition in 1901 (Wilson's Oklahoma Statutes 1901; Oklahoma Statutes 1915; Dillow 1981; *The Oklahoman* 1905a, 1905b, 1906, 1907, 1914). The American Tobacco Company monopoly vigorously resisted this cigarette prohibition (*The Oklahoman* 1905a, 1905b, 1906, 1907, 1914). By 1914, enforcement of the cigarette ban was no longer occurring. The reasons were summarized in an April 1914 newspaper story in the statewide newspaper *The Oklahoman*: "Oklahoma now has a law against selling or giving away cigarettes or cigarette papers, but while this is true the law has been ignored, and on the street corners of every city town and hamlet in the state boys can be seen publicly smoking cigarettes at all times" (*The Oklahoman* 1914). In 1915, the Oklahoma legislature amended the cigarette prohibition law to prohibit adults from providing cigarettes or cigarette papers to minors. This in effect ended cigarette prohibition (Oklahoma Statutes 1915).

ANTI-TOBACCO ADVOCACY FROM EARLY 1990S TO 2001

Between 1915 and the early 2000s the tobacco industry maintained a complete stronghold in the state, including the adoption in 1987 of smoker-friendly legislation requiring restaurants to provide smoking sections and preempting any local anti-smoking ordinances. Big Tobacco retained former state legislators as lobbyists, correctly predicting that their knowledge of the inner workings would result in complete hegemony. These tobacco lobbyists ran an efficient, well-oiled system of control that extended far beyond the capitol, not only upholding but catering to the rights of smokers from Enid to Idabel, Muskogee to Lawton (Spivak and Givel 2005).

In the early 1990s, the Oklahoma Alliance on Health or Tobacco (OAHT), known as the Tobacco Free Oklahoma Coalition (TFOC) after 2000, was organized between members of various government, health care, and nonprofit organizations to attempt to address tobacco issues and lobby on behalf of health interests (Oklahoma Alliance on Health or Tobacco 2004a, 2004b). Included in the Alliance were the Oklahoma affiliates of the American Lung Association, American Cancer Society, American Heart Association, Oklahoma State Medical Association, and the Oklahoma State Department of Health.

Into the mid-1990s, under director Alice McGrew, the Tobacco Free Oklahoma Coalition attempted to counter the "status quo" on tobacco issues at the state capitol, where they observed that

> Tobacco industry lobbyists considered Oklahoma to be a "love fest" when it comes to support for pro-tobacco, anti-health legislation . . . most legislative efforts initiated or endorsed by the Coalition have failed to become law while

> policies advocated by the tobacco industry have largely been adopted, despite Coalition opposition. Proposed tobacco prevention and control legislation has often been assigned to unfavorable legislative committees where it simply has not been heard. (McGrew 1996)

Richard Barnes, an attorney and public relations manager for the American Lung Association of Oklahoma, in confirmation of McGrew's assessment, also stated: "Oh, there was plenty of activity (by organized health groups), there was just no (significant) success" (Barnes 2004b).

Nonetheless, the OAHT began to make some limited progress in the early and mid-1990s. A 1992 joint resolution passed the state legislature that prohibited tobacco use on the premises of licensed day care centers. The following year, a Senate resolution congratulated the president of the University of Oklahoma for proposing a smoke-free campus (which did not officially happen until 2012), and in 1994, House Bill 2129 amended the weak 1987 Smoking in Public Places Act, finally allowing restaurants to establish smoke-free facilities; they were previously required to have a smoking section (Barnes 2004a; McGrew 1996). Despite these advances, McGrew was keenly aware of the inept state of health advocacy in this early period.

> I worked for the American Cancer Society . . . and always during that time [early 1990s] I worked in what we call Public Education . . . and which was at that time probably made up of Heart, Lung, Cancer, (American Heart Association, American Lung Association, and American Cancer Society) and the Health Department . . . just a very small group . . . and one of my roles was lobbying . . . we would go to the capital and always went in the name of Heart, Lung, and Cancer. I mean this is just what we did, and we would have little information sheets, and believe me. We did not know what we were doing. (McGrew 2004)

The OAHT saw even the 1995 failure of House Bill 1007 as a measure of progress. The bill would have allowed smoke-free policies in publicly owned buildings across the state. The Alliance supported the bill strenuously enough to pass the House of Representatives and be assigned a Senate committee that would hear it, a substantial political achievement for health advocates at the time. However, "due to vigorous opposition by the tobacco lobbyists," it was narrowly defeated (McGrew 2004).

The ineffectiveness of the Alliance's lobbying efforts during these times may have been the result of health lobbyists' compromising approach to tobacco legislation. Barbara Kumpe, Advocacy Director for the American Heart Association in Arkansas and Oklahoma, recalled how the health community shied from the "all or nothing" aggressiveness occasionally exhibited by the Oklahoma State Department of Health, and preferred to work in tandem with the industry.

> I do believe that when you work with issues like the tobacco issue or nursing home issues if its an issue, I think you have to know what your opposition is about, I think you have to be able to negotiate with your opposition. I don't think you that you go "I'm not ever speaking to you I don't want to talk to you, I'm against you," you can't lobby that way. I mean that is my perception you have to know, you have to be able to talk, you have to negotiate things and I think with the tobacco issue and a lot of issues you have to take little steps. And some people want all or nothing, I would rather have something to get started with than to have absolutely nothing, and I have seen that in some of the health department approaches. (Kumpe 2004)

Kumpe's assertion, that health interests needed to be willing to work with tobacco lobbyists, personified the insider approach.

Bob Miner, chair of the OAHT from 1997 through 1999, also remembered the 1990s as a transitional period of early organization and growing pains.

> I would classify the state of public health advocacy in this early period as getting acquainted, exposing more personnel to the advocacy and legislative arena, beginning education of policy makers as well as the general public on tobacco issues, and beginning to cultivate both inter-organization communications through a coalition and grass-roots organizations to communicate with policy makers in the future. (Miner, 2004)

Miner also noted that during this period, there were some special accomplishments. These included

> funding for the $1.2 million, three-year Smokeless Oklahoma program. Lauching NATCO, the Native American Tobacco Coalition of Oklahoma. Rallies at the capitol for at least three successive years (some sponsored by ACS and AHA), including one with more than 1,500 students and many celebrities. Regional grassroots workshops boosting clean indoor air legislation throughout the state. The first paid lobbyists for the Coalition, per se. A fund raising dinner with Joe Garagiola at the Cowboy Hall of Fame, to help fund the first teen summit and full-scale launch of Students Working Against Tobacco (SWAT). Development of the first State (Tobacco Control) Plan. A lot of growth. Increased capitol presence, recognition and respect. A very few limited legislative successes, but no great breakthroughs yet. (Miner 2004)

STATE PREEMPTION OF LOCAL CLEAN INDOOR AIR ORDINANCES

The issue of preemption by the state of stronger clean indoor air laws provides a primary example of the political clout of the tobacco industry and its allies prior to 2001, with a legislative legacy that has lasted until 2012. Following the 1986 Surgeon General's report on secondhand tobacco smoke, some employers in Tulsa, under the leadership of a local dentist working with the American Lung Association of Oklahoma, began to openly favor smoke-free workplaces, prompting opponents to suggest enacting a state law on smoking in public places. Senate Bill 327 began the 1987 legislative session with impossibly strong language for the standards of the time, prohibiting smoking in all public places, including bars, and proposing stiff enforcement with misdemeanor criminal fines. By the time SB327, called the *Smoking in Public Places Act*, passed the legislature and was signed by Republican Governor Henry Bellmon in June, it was a completely different bill. Richard Barnes, government relations manager for the American Lung Association of Oklahoma, refers to the law as the "Unclean Indoor Air Act of 1987," and recalled

> It no longer prohibited smoking except in designated smoking areas meeting minimum standards, but merely provided that "smoking and nonsmoking areas may be designated" in a public place. Smoking was prohibited only in a designated nonsmoking area. . . . Also stripped of the bill were all of the enforcement provisions and all of the minimum standards for protecting nonsmokers except for enforcing no smoking in the nonsmoking areas by asking smokers to refrain from smoking in nonsmoking areas upon request of a client or employee suffering discomfort from the smoke. Worst of all was the addition of the dreaded "preemption" section that prohibited cities and towns from enacting laws that are "more stringent than those of this act." As a result of the different versions passed in each house, SB327 was referred to a Conference Committee. What came out of conference was worse yet. It now mandated the designation of smoking and nonsmoking areas . . . and the dreaded "preemption" section was still intact. (Barnes 2001b)

By at least 1992, the industry was becoming alarmed at the potential impact of workplace smoking restrictions. A Philip Morris analysis in that year stated:

> Total prohibition of smoking in the workplace strongly affects industry volume. Smokers facing these restrictions consume 11%–15% less than average and quit at a rate that is 84% higher than average. Only 6.4%–10.3% of smokers face total workplace prohibition but these restrictions are rapidly becoming more common.
>
> Milder workplace restrictions, such as smoking only designated areas, have much less impact on quitting rates and very little effect on consumption.
>
> If smoking were banned in all workplaces, the industry's average consumption would decline 8.7%–10.1% from 1991 levels and the quitting rate would increase 74% (e.g., from 2.5% to 4.4%). (Philip Morris 1992)

In order to ensure that any authority to pass smoking restrictions would be confined to the state legislature, where the industry could centralize its lobbying efforts, preemption clauses prohibited local municipalities from enacting tobacco control ordinances more stringent than the state statutes (American Cancer Society 1998). A 1985 R. J. Reynolds state analysis document noted that smoking regulation at the local level had yet to appear in Oklahoma, but that anti-tobacco forces may take such a strategy: "Affiliates of the American Cancer Society and the American Lung Association have proposed legislation at the state level, but no attempt has been made to enact a local smoking control law. However, such an effort is probably inevitable and will likely come in Tulsa, where some members of the medical community have made public remarks concerning the need for such a law" (R. J. Reynolds 1990).

When the 1987 Smoking in Public Places Act (SB329) was passed, the Tobacco Institute noted, "A very mild smoking restriction law was passed in 1987 which included language preventing municipalities from regulating smoking" (Tobacco Institute 1990a). The tobacco industry continued to vigorously support preemption throughout the 1980s and 1990s.

The Tobacco Institute demonstrated its commitment to preemption of smoking ordinances in a 1990 memorandum which began with the announcement that "As part of the industries [sic] proactive program, we have introduced in Oklahoma HB 1812, a hiring discrimination bill" (Tobacco Institute 1990b). The bill had been referred to the House Business and Commerce Committee and was ordinarily something to which the industry would have been favorable. However, the Tobacco Institute quickly realized where its priorities lay.

> Ken Nance has expressed grave reservations about trying to move the bill . . . Ken's concerns are: an attempt by our opponents to amend the bill to remove the local pre-emption on the state's smoking restriction law (We all agree-maintaining that pre-emption is top priority Oklahoma.). . . . Opinion of the majority of industry lobbyists in Oklahoma City is that while a hiring discrimination bill would be helpful to the industry, the chance of jeopardizing the local pre-emption must outweigh all other concerns. (Tobacco Institute 1990b)

While no state had managed to reverse preemption at the time, the Coalition decided to attempt this goal in November 1995. After organizing an interim study on the issue by the House of Representatives, resulting in substantial media attention, the House proposed HB 2493 in January 1996 with 44 House and 14 Senate coauthors. A similar bill had passed the House the year before but stalled in the Senate. During the media coverage, Oklahoma Health Commissioner Jerry Nida explained an important benefit to preemption repeal: better enforcement for Synar, the youth access restriction that prohibits the sale of tobacco to minors.

Prior to 1997, the Oklahoma Alcoholic Beverage Law Enforcement Commission was unable to effectively enforce youth access restrictions (McGrew 1996). Seven cities had already adopted resolutions requesting that the legislature allow them to impose their own tobacco ordinances: Ada, Claremore, Oklahoma City, Midwest City, Stillwater, Warr Acres, and Weatherford (*Tulsa World* 1996a). A *Tulsa World* article cited former New York Yankees member Bobby Murcer and representatives of the American Cancer Society as urging the legislature to support the bill (*Tulsa World* 1996b). Even Governor Keating indicated that he would sign the bill if it passed the legislature, which would have completely countered the language in state tobacco laws with regard to preemption (McGrew 1996). The language of the proposed bill was revolutionary:

> Nothing in the Smoking in Public Places Act shall be construed to supersede or in any manner affect a municipal smoking ordinance provided that the provisions of such ordinance shall be the same as or more stringent than those provided for in the Smoking in Public Places Act, including the enforcement provisions . . . Nothing in the Prevention of Youth Access to Tobacco Act shall be construed to supersede or in any manner affect a municipal tobacco ordinance provided that the provision of such ordinance shall be the same as or more stringent than those provided for in the Prevention of Youth Access to Tobacco Act. (State of Oklahoma 1996)

The tobacco lobby enlisted the aid of the Oklahoma Restaurant Association and the Oklahoma Grocers Association to fight the newly proposed anti-preemption law (McGrew 1996). Philip Morris sent a notice to retailers in February 1996 that began with the following announcement:

> The Ada, Ok. City Council today passed a new city ordinance that imposes sweeping new regulations, stiff fines and penalties for retailers that sell cigarettes and tobacco products from their stores within the Ada city limits. . . . The Ada City Council is also considering enacting rules and regulations that would require retailers to record the name, age, type of product purchased, etc. for each customer purchase of any tobacco product. . . . A new ordinance to ban smoking in all public places will be considered at the next city council meeting. (Philip Morris 1996a)

Of course, the Ada City Council was planning no such ordinance. The announcement was an attention getter. The Philip Morris notice to retailers went on to state that:

> What you have just read is fiction . . . for now. Legislation has just been introduced into the Oklahoma House of Representatives called House Bill 2493 that will repeal the preemption clauses found in the "Smoking in Public Places

> Act" and the "Prevention of Youth Access to Tobacco Act" . . . It means that H.B. 2493 as introduced by Rep. Raymond L. Vaughn, Jr., will repeal the pre-emption clauses in the Oklahoma Tobacco Laws which opens the door for any town or city or county to impose stricter or harsher regulations, fines, penalties, ordinances, etc. . . .
>
> Can H.B. 2493 be stopped? Absolutely! A flood of calls from angry store keepers will bring this nonsense to a screeching halt . . . Your senator and representative will take your opinion seriously . . . explain the impact of increased regulation and red tape and the un-reasonable burden that this will place on your livelihood. (Philip Morris 1996a)

Both the restaurant and grocer associations went into action. The Oklahoma Restaurant Association sent an official statement to the Oklahoma Newswire in January 1996, declaring that "House Bill 2493 will weaken Oklahoma's smoking laws. This was to occur by confusing and harassing the public and make [sic] it harder and a lot more expensive for business to educate their employees about the existing laws" (Philip Morris 1996b). The statement urged customers and businesses to contact their representatives and "urge them to fight the bill for the good of Oklahomans" (Philip Morris 1996b).

Clarity and uniformity were clearly not the primary concern for the industry or their supporters. Philip Morris' 1996 tract titled "In Opposition to Efforts to Repeal Oklahoma's Uniform Smoking and Tobacco Sales Laws" adds to the rhetoric about confusion and then comes more to the point.

> Some localities may do nothing, while others may decide to ban smoking. As smokers (who may account for 25% of patrons, or more) seek out accommodating establishments, restaurants in communities that prohibit smoking could lose significant revenues . . . that the current Oklahoma law makes sense is further reinforced by a number of recent studies indicating that a clear majority of the general public prefer the accommodation of nonsmokers and smokers instead of a ban. (Philip Morris 1996c)

Clearly, the industry was not concerned with helping businesses avoid confusion and address smoking "uniformly and effectively." Rather the industry was concerned with making sure that the legislature enacted limited and ineffective state clean indoor air regulation, which could not be strengthened anywhere in the state by local governments.

REPRESENTATIVE RAY VAUGHN

In 1996, Rep. Ray Vaughn (R.-Edmond), a long-time proponent of tobacco control, was also pressured by his colleagues in the House Leadership to withdraw his own bill before a committee vote, limiting the passing legislation to one that only allowed municipalities to mirror state law (eliminating "super-preemption," a condition disallowing municipalities from enacting identical legislation as the state, so that such laws can be locally enforced). Doug Matheny and Bob Miner discussed how the defeat occurred.

> The anecdotes that I heard were about a meeting in January 1996 between Rep. Ray Vaughn and certain individuals who went into an office with the door closed and came out with an agreement that Rep. Vaughn would withdraw his bill to repeal preemption and that his companion bill would be supported, providing it only allowed localities to pass ordinances on clean indoor air and prevention of youth access that mirrored the provisions of state laws and were no stronger than state laws. The reports I heard identified the people who met with Rep. Vaughn as being tobacco lobbyists. The details of this event and the negotiations that ended this very serious attempt to repeal preemption have never been out in the light where I could see them. (Matheny and Miner 2004)

Philip Morris's Draft Oklahoma State Plan for 1997 recalled the legislative victory with a note of caution, stating, "Oklahoma currently has sales/marketing preemption. In '96 strong attempts failed to reverse this and expect similar attempts in '97" (Philip Morris 1997).

Since Ray Vaughn's humiliation in 1996, preemption has sat sturdily in Oklahoma state law without serious challenges. The 1997 Draft Plan, despite its cautionary reference to future attempts at repeal, noted a somewhat favorable landscape. "The political environment is somewhat favorable. The legislature remains balanced . . . Speaker Glen Johnson has abdicated his position to Lloyd Benson who supports preemption, but generally not with us on other issues" (Philip Morris 1997: 1).

In the spring of 2002 legislative session, Representative Vaughn attempted again to introduce legislation repealing preemption (Ford 2002a). The government relations manager for the American Lung Association, Richard Barnes, even announced publicly that preemption was unconstitutional, and that the Oklahoma constitution's "constitutional home rule" provision for cities like Tulsa guarantees that they may "pass ordinances that may conflict with state law as long as that ordinance is local in nature" (Ford 2001). Tulsa never passed such an ordinance, and Vaughn never saw Oklahoma's preemption clauses repealed.

Chapter Three

The Influence of the Tobacco Industry Lobby

This chapter examines industry lobbying efforts in Oklahoma from the late 1980s until the early 2000s, noting the associations of specific lobbyists with their clients, as well as the activities and tactics employed over the years. Descriptions of major lobbyists, their primary clients and expenditures, and verbatim accounts of their activities follow an overview of the early industry influence.

TOBACCO INDUSTRY LOBBYING BEFORE 2001

Early tobacco lobbying efforts during the late 1980s and early 1990s were hailed as generally successful and unencumbered. A 1989 Philip Morris report, in summary of southwestern states' political outlooks under the heading Lobbyists, noted that:

> In OK, [Oklahoma] the industry lobby team is absolutely superb and seems to be able to hold leadership to coming out in the press against a cigarette tax every time the Governor brings it up, which is often. All the lobbyists were former members and just do a great job. So our defensive strategy is just to keep on keeping on. We use events, charitable contributions, and entertainment of standing committees. Not much caucus stuff here. A major improvement is we have to establish an Oklahoma PAC or give Tom Rogers a raise. So far we've done fine without it but it sure would help out. Tom is sending me the stuff. (Philip Morris, 1989)

The same document revealed the industry's confidence regarding the cooperation of allies on tobacco-related legislative issues, briefly summarizing the outlook in Oklahoma under the heading of Constituents.

> Again, a love fest. They are all best friends. We sponsored a legislative bar-b-que [sic] with TI, Tobacco Wholesalers and RJR earlier this year. PM works with the restaurant associations, nursing home associations in opposing legislation to remove the state preemption of municipal ordinance banning smoking. Groups we will be working on will be state employee associations, teacher associations, and similar groups for smoker's rights. (Philip Morris, 1989)

A 1990 Tobacco Institute memo addressing tax initiatives also stressed the industry's recognition of its successful influence of their past lobbying efforts, and their reliance on continuing cooperation in the future.

> Essential to success in Oklahoma . . . will be a continuation of the close and cooperative efforts of all industry lobbyists, as led by Tobacco Institute legislative consultant Ken Nance. In addition, member company lists may be needed to generate grassroots contacts with legislators. (Tobacco Institute1990c)

The same year, Philip Morris constructed a short list of "Friendliest Legislators" and "Least Friendly Legislators," as well as "Allied Groups," including the Oklahoma Association of Tobacco Distributors, the Oklahoma Restaurant Association, and the Oklahoma Retail Grocers Association, and "Adversary Groups," which included the Oklahoma Cancer Society and the Oklahoma Lung Association (Philip Morris 1989). A Tobacco Institute memo in 1992 further describes the general atmosphere of confidence in lobbying activities.

> Industry success in Oklahoma illustrates the benefits of a coordinated and well-planned effort. The entire Oklahoma tobacco team, composed of Ken Nance, TI Legislative Counsel, Mandell Matheson, RJ Reynolds Legislative Counsel, Tommy Rogers, Philip [sic] Morris Legislative Counsel and Richard Huddleston, Smokeless Tobacco Council and the Wholesalers Legislative Counsel are to be commended for their fine work. (Tobacco Institute, 1992c)

Ronald Morris, Vice President of Tobacco Institute's State Activities Division, wrote to Mary Thurber, executive director of the Oklahoma Association of Tobacco & Candy Distributors, "The effort you outlined . . . is just what we want to see, wholesalers protecting their business at the state, local, and federal levels. You have an active, well directed organization in Oklahoma" (Tobacco Institute, 1995b).

TOBACCO INDUSTRY LOBBYING FROM 2001 TO 2003

The historic effect of tobacco lobbying in Oklahoma is largely the result of the talent and professional effort of specific lobbyists, who have become familiar and notorious persons among the public health community. Below is a brief description of the most prominent lobbyists in the state and whom they represented.

Richard L. Huddleston

Huddleston was registered as a lobbyist representing the Oklahoma Association of Tobacco and Candy Distributors and U.S. Tobacco (Smokeless Tobacco) in 1999–2000, and U.S. Tobacco in 2001–2002. His only gift report attributed to a tobacco interest was for a small contribution to Rep. Ferguson on behalf of "U.S. Smokeless" (Oklahoma Ethics Commission, 2001a; Oklahoma Ethics Commission, 1999).

Mandell Matheson

Mandell Matheson agreed to a rate of one hundred dollars per hour in his contract with R. J. Reynolds in September 1988. The terms specified by RJR in the agreement were for "Services and assistance necessary and appropriate to your direct representation of Reynolds before the various regulatory, administrative or legislative bodies in the State of Oklahoma, and . . . monitoring and surveillance of regulatory and legislative activities with potential impact on Reynolds" (R. J. Reynolds, 1988a). Matheson also lived up to his employer's expectations, providing valuable information and counsel to R. J. Reynolds. In 1994, an RJR internal memo noted Matheson's recent discovery that "challenger Ms. Debbie Blackburn (running for open seat) has 'changed her tune' and now is on the bandwagon as a staunch anti-tobacco foe!" The company's response to their lobbyist's timely disclosure was swift. "Therefore, please CANCEL [capitals original] our earlier request to issue her a political contribution in the amount of $100 . . . the checks have not been cut yet and Mandell thought we could save ourselves some embarrassment by canceling her off our 'friend list' ASAP" (R. J. Reynolds, 1994).

In the 1995 Oklahoma 2nd Congressional District race, RJR found it could not make a direct contribution to Glen Johnson's campaign against the incumbent, Rep. Tom Coburn. Working with an RJR Vice President, Tommy Payne, Matheson facilitated:

> A way through a third party that we could get involved . . . if we can work it. There is going to be a major effort to repeal our preemption in the 1996 session and we will need Speaker Johnson's help if we are to be successful in stopping this. His record on our issues has been good and I would like to see you help him out if possible. (R. J. Reynolds, 1995a)

A few months later, an email memo sent to six RJR employees expressed thanks "for contributing to the congressional race of Speaker Glen Johnson in Oklahoma." The memo goes on to recall "the speaker was quite helpful to us in our battle to fight the repeal of preemption . . . a battle in which we were successful. I know Speaker Johnson appreciates it, but not nearly as much as Mandell Matheson or I do" (R. J. Reynolds, 1995b). Also, in 1995, when the Oklahoma Legislature was contemplating a repeal of preemption for clean indoor-air and youth access preemption, Matheson helped to implement a successful "grassroots program" involving direct connect phone calls, connecting Oklahomans to legislators on the Commerce, Industry and Labor Committee and to the Speaker of the House (R. J. Reynolds, 1995c). Gift reports to Oklahoma elected officials from 2000 to 2003 by Matheson indicate a number of contributions on behalf or RJR (table 3.1) (Oklahoma Ethics Commission, 2001b).

Kenneth Nance

In 1992, the Tobacco Institute, and hence the industry, had good reason to hire Nance as a contract lobbyist; he was clearly worth every penny. An in-

Table 3.1. Lobbyist or Other Person Gift Report (January 17, 2001–January 15, 2003).

State Rep. Tad Jones	07-29-2000	Green Fees	125.00
State Rep. Greg Piatt	07-29-2000	Green fees	125.00
State Sen. Grover Campbell	07-29-2000	Green Fees	125.00
State Sen. Jim Dunlap	12-15-2000	Food Box	37.76
Rep. David Braddock	08-08-2000	Green Fees	75.00
Sen. Robert Milacek	08-08-2000	Green Fees	75.00
Sen. Jeff Rabon	08-08-2000	Green Fees	75.00
State Sen. Robert Milacek	07-17-2001	Golf Fees, Cart, Green Fees	150.00
State Rep. Ron Peters	07-17-2001	Golf Fees, Cart, Green Fees	150.00
State Sen. Bruce Price	07-17-2001	Golf Fees, Cart, Green Fees	150.00
State Rep. Ron Peterson	08-04-2001	Golf Fees, Lunch, Golf Gift	175.00
State Rep. John Smaligo	08-04-2001	Golf Fees, Lunch, Golf Gift	175.00
Sen. Jim Dunlap	08-14-1999	Golf Cart, Fees, and Green Fees	100.00
Sen. Grover Campbell	07-20-1999	Golf Cart, Fees, and Green Fees	75.00
Sen. Jeff Rabon	08-06-2002	Golf, Green Fees	125.00
Rep. Ron Peters	08-10-2002	Golf, Green Fees	125.00
Rep. Kris Steele	08-10-2002	Golf, Green Fees	125.00
Rep. Ron Peters	08-12-2003	Golf Course and Cart Fees	100.00
Sen. Jim Dunlap	08-02-2003	Golf Course and Cart Fees	125.00
Rep. Ron Peterson	08-02-2003	Golf Course and Cart Fees	125.00

Source: Oklahoma Ethics Commission, Form L-2

ternal Tobacco Institute memo reporting the 1992 legislative session wrap-up for Oklahoma lauded Nance's defeat of Governor Walters' proposed 5 cent per pack tax increase:

> To counter this expected thrust, the "Sooner State" tobacco-team continued to operate in a coordinated effort . . . The team, under the skilled and capable leadership of Ken Nance, TI Oklahoma Legislative Counsel, worked closely with members of the legislature, on both sides of the aisle, and by the time that Governor Walter's proposal . . . arrived in April 1992, the tax was literally pronounced "dead on arrival" in the Oklahoma House of Representatives. (Tobacco Institute 1992a)

The month before this memo Nance had advised the Institute that it could lawfully underwrite a portion of the 1992 Legislative Reception by making a $500 purchase at Byron's Liquors in Oklahoma City, the specific liquor purchases to be selected by Nance. Tobacco Institute found the expense worthwhile, since "this reception is of a general nature and will involve many members of the Oklahoma Legislature" (Tobacco Institute 1992b).

In 1994, the Tobacco Institute, which was represented by Nance, wrote a glowing evaluation of his work, stating that:

> Ken Nance has done an outstanding job for the Tobacco Institute. Under his leadership, pre-emption was passed in 1986 and privacy was passed in 1991. For the past seven years, under both Republican and Democratic governors, proposed tobacco-tax increases have been repeatedly defeated. Nance and the other members of the tobacco team in Oklahoma work extremely well together and appear to have a considerable amount of strength in the political process there. (Tobacco Institute 1994)

When the Oklahoma Department of Labor attempted to enact legislation in 1994 that would have restricted smoking in government buildings to separately ventilated areas, the Tobacco Institute noted that Nance, "responded quickly and decisively and through his efforts we were able to delay any action on this issue indefinitely" (Tobacco Institute, 1994, 1995a).

Ken Nance would be offered $47,000 to provide services during the course of 1996, which as specified in the agreement letter would include: "To represent The Institute before state legislative, regulatory and administrative bodies on all issues of interest to the tobacco industry; to report to The Institute on all such legislative, regulatory or administrative actions, including bill introductions, hearings and votes, and proposed regulations and other administrative actions" (Tobacco Institute 1995b).

The language in this 1995 lobbyist agreement (above) makes clear that Nance was not just representing a particular company's interests in

Table 3.2. Lobbyist or Other Person Gift Report, Attachment (July 18, 2002).

Rep. Fred Morgan	04-01-2002	Dinner – Lorillard Tobacco	78.25
Rep. Richard Philips	04-03-2002	Dinner – Lorillard Tobacco	78.25
Sen. Glenn Coffee	04-03-2002	Dinner – Lorillard Tobacco	78.25
Sen. Brooks Douglass	04-03-2002	Dinner – Lorillard Tobacco	78.25
Sen. Jim Dunlap	04-03-2002	Dinner – Lorillard Tobacco	78.25

Source: Oklahoma Ethics Commission, Form L-2.

Oklahoma legislative and regulatory matters, but rather the interests of the entire "industry." The author of the agreement, Tobacco Institute president Samuel Chilcote, in using the general term "industry" rather than just the name of his own organization, conveyed an understanding that the political goals that their lobbyist would pursue were part of a policy agenda common to tobacco interests in general.

In the 2003–2004 period, he was listed as "monitor only" for Brown & Williamson, in addition to numerous non-tobacco clients (Oklahoma Ethics Commission 1999a). On January 1, 2004 he terminated his representation of Brown & Williamson (Oklahoma Ethics Commission 2004g). His only reported gifts to Oklahoma elected officials on behalf of a tobacco principal were as shown in table 3.2 (Oklahoma Ethics Commission 2002).

Tom Rogers

In 1996, the executive director of the Oklahoma Association of Tobacco & Candy Distributors, Mary Thurber, wrote to the Regional Director of Government Affairs, Ed Beauchemin, that: "The members of this association have worked closely in recent weeks with our legislative consultant, Richard Huddleston, and your lobbyist in Oklahoma, Tom Rogers, to effect the demise of HB 2493 and retain the pre-emption clause in the Oklahoma statutes" (Philip Morris 1996d).

Rogers represented Philip Morris, sometimes by its own name, sometimes under the name Altria Corporate Services, Inc. from at least 1999 through 2004 (Oklahoma Ethics Commission 1999c). No gift reports are on file.

Benny Vanatta

Vanatta represented the Oklahoma Restaurant Association (ORA), an industry ally in Oklahoma, and the Smokeless Tobacco Council from at least 1999 (the earliest year for which lobbyist registration is available) through 2004 (Oklahoma Ethics Commission 1999b). No gift reports are on file. As described in the next chapter, Vanatta and the ORA would play a prominent role in fighting the clean indoor air campaign in the early 2000s.

ALLIED ORGANIZATIONS

The warm relationship between the industry and allied organizations is exemplified in a 1995 letter from Tobacco Institute's State Activities Division Vice President, Ronald Morris, to Mary Thurber, executive director of the Oklahoma Association of Tobacco & Candy Distributors: "Your comments concerning the capabilities of your organization and your lobbyist are understated. I have known many of your members since the mid-seventies, and have worked with Richard Huddleston since he was Clerk of the House. We have never lacked the support of the wholesaler organization in Oklahoma" (Tobacco Institute 1995b).

Throughout the 1990s, and during the successful tobacco control campaigns in the past few years, industry efforts have been supported by organizations such as the Oklahoma Association of Tobacco & Candy Distributors, the Oklahoma Restaurant Association, the Oklahoma Grocers Association, and the Oklahoma Retailer's Alliance. As will be delineated in subsequent chapters of this book, these organizations have lobbied and publicly campaigned against tobacco control in conjunction with industry interests (Barnes 2004a; Carter 2004h; Oklahoma Ethics Commission 2004f; Philip Morris 1989, 1999a; R. J. Reynolds 1990; Talley 2002b; Associated Press 2004).

CAMPAIGN CONTRIBUTIONS

According to the available records archived at the Oklahoma Ethics Commission, the total amount contributed by the industry to individual legislators, over the course of the period 1991–2004, was $96,700 (table 3.3). The average amount per legislator was $575.60 over the entire period. Unsurprisingly, legislators who served a larger number of terms tended to receive larger total contributions. Governor Keating received tobacco funds only during one legislative period, totaling $5,000 during 1997–1998, half from each of Philip Morris and R. J. Reynolds.

TOBACCO LOBBY ADVOCACY FROM 2001 TO 2003

Between 2001 and 2003, the prior political hegemony that the tobacco lobby maintained and enjoyed in Oklahoma began to dwindle. In March 2001, the Tobacco Merchants Association noted with respect to a vote on repealing stronger state preemption of local clean indoor air ordinances: "The Oklahoma Senate voted 25–18 for bill 752 that authorizes local governments to overstep a State law on smoking by imposing stricter ordinances. Senator Ben

Table 3.3. Contributions to Leadership of the Forty-Ninth (2003–2004) Oklahoma Legislature, 1995–2004.

Name	*Title*	*Party*	*House*	*OATCD*	*PM*	*RJR*	*UST*
Don Armes	Majority Whip	R	House	$0	$100	$250	$0
Jari Askins	Democrat Leader	D	House	$0	$0	$0	$0
Gus Blackwell	Majority Whip	R	House	$0	$100	$250	$0
David Braddock	Deputy Democrat Floor Leader	D	House	$0	$600	$100	$550
Lance Cargill	Majority Floor Leader	R	House	$0	$200	$0	$100
Dale DeWitt	Majority Whip	R	House	$0	$0	$0	$0
Todd Hiett	Speaker	R	House	$100	$300	$200	$1,050
Pam Peterson	Majority Whip	R	House	$0	$0	$0	$0
Joe Sweeden	Democrat Whip	D	House	$0	$400	$450	$0
Opio Toure	Democrat Floor Leader	D	House	$0	$100	$200	$0
Dale Turner	Deputy Democrat Floor Leader	D	House	$0	$60	$100	$0
Susan Winchester	Speaker Pro Tempore	R	House	$0	$0	$0	$0
Randy Brogdon	Minority Whip	R	Senate	$0	$0	$0	$0
Glen Coffee	Minority Floor Leader	R	Senate	$0	$250	$0	$300
Ted Fisher	Majority Floor Leader	D	Senate	$0	$850	$200	$700
Mike Morgan	President Pro Tempore	D	Senate	$0	$0	$250	$400
Susan Paddack	Majority Whip	D	Senate	$0	$0	$0	$0
Nancy Riley	Minority Whip	R	Senate	$0	$250	$0	$0
Total				$100	$3750	$1800	$3,100

Note: The earliest any of these persons held office was the 1995–1996 Forty-Fifth Legislature.

Source: Oklahoma Ethics Commission.

Robinson (D-Muskogee) expressed surprise at the tobacco lobby's failure to vote the bill down, while former mayor Sen. Dave Herbert (D-Midwest City) said the new law was unnecessary. The Oklahoma Restaurant Association has also opposed the bill that is yet to be approved by the House" (US Tobacco Weekly 1998).

The Oklahoma House never passed the preemption law but the fact that it got through the Senate was unprecedented compared to the political atmosphere in the 1990s. Another December 2001 R. J. Reynolds internal document provided talking points to oppose stronger clean indoor air legislation embodied in SJR 21, including accommodation of smokers and non-smokers in the same public areas, and no link between secondhand tobacco smoke and lung cancer, heart disease, and other diseases (R. J. Reynolds 2001a). A 2001 Philip Morris memo on the political outlook in Oklahoma for 2002 also noted that: "Falling oil and gas revenues, coupled with events connected to the September 11 tragedy, have created an unstable economic environment. A likely source of revenue in 2002 would be a tax increase on tobacco" (US Tobacco Weekly 1998). A tobacco tax increase did not pass in 2002, but instead was enacted with the approval of Oklahoma voters in 2004.

Meanwhile, as expected, the industry was quietly monitoring with growing concern the ongoing clean indoor air campaign that the new health commissioner, Dr. Leslie Beitsch, was waging with support from Oklahoma health groups (Philip Morris 2001; US Tobacco Weekly 2002). This campaign included a well-publicized news conference in Tulsa, where the Tobacco Free Oklahoma Coalition announced that one of the traditional tobacco allies, Mazzio's Corporation, had broken ranks and was implementing a smoke-free policy in all 93 Mazzio restaurants in Oklahoma. Another internal Philip Morris document in February 2002 worried that Beitsch had publicly called for a one dollar a cigarette pack tax increase (Borelli 2002).

The nearly complete hegemony that tobacco lobbyists had enjoyed for decades in Oklahoma had thus finally begun to decline, culminating in SJR 21, the first effective clean indoor air bill, which passed in 2003. Despite private worries behind-the-scenes and lobbying to terminate it, the bill passed with an ongoing campaign by the Oklahoma Health Commissioner, with assistance from organized health organizations. Their applied outsider public pressure on politicians, holding them publicly responsible for their pro-tobacco actions. On the part of the Oklahoma Health Commissioner, these outsider advocacy actions represented public administration ethics that forwarded the public interest, representing compassion, benevolence, fairness, and optimism in countering the significant public health impacts of tobacco use. The story of this new approach and its political triumph is the subject of the next chapter.

Chapter Four

Leslie Beitsch and the 2001–2003 Clean Indoor Air Campaign

Prior to 2001, tobacco industry efforts in Oklahoma were largely triumphant. One key example of this success occurred in 1986 when the American Lung Association, reflecting an insider approach, initiated an effort in Tulsa to enact a local secondhand smoke ordinance, tapping local dentist Dr. Dan Brannin to spearhead the campaign. In reaction to this effort in Tulsa, opponents of local secondhand tobacco smoke legislation went to the Oklahoma legislature in 1987 and lobbied to pass the *Smoking in Public Places Act*, a pro-tobacco industry bill, also known as Senate Bill 327 (Smoking in Public Places Act 1987). Despite these early efforts, Oklahoma remained squarely under the control of tobacco interests.

Senate Bill 327 began the 1987 legislative session with impossibly strong language for the standards of the time, prohibiting smoking in all public places, including bars, and proposing stiff enforcement with misdemeanor fines. By the time the bill passed the legislature and received Governor Henry Bellmon's signature in June, it only required the designation of smoking and non-smoking areas in enclosed indoor public areas, including state and local government buildings, educational and health facilities, museums, theaters and performing arts centers, and restaurants (Smoking in Public Places Act 1987). Jails, enclosed or separate bars, and facilities for private functions not under the control of state or local government were exempt (Smoking in Public Places Act 1987). Section 4 (A) of the law stated that "Existing physical barriers and ventilation systems shall be used to minimize smoke in both smoking and adjacent nonsmoking areas," and stipulated that owners and operators of public places consisting of a single room "shall be in compliance with this act if an area of the room is reserved and posted as a nonsmoking area" (Smoking in Public Places Act 1987).

The new bill's weak enforcement provisions included requiring the posting of a sign stating that smoking in designated non-smoking areas was prohibited and asking smokers to refrain from smoking in nonsmoking areas if a client or employee suffers discomfort or hypersensitivity from the smoke (Smoking in Public Places Act 1987). Additionally, the bill contained a provision preempting local governments from enacting legislation more stringent than the state law (Smoking in Public Places Act 1987). Richard Barnes, government relations manager and attorney for the American Lung Association of Oklahoma, referred to the law as the "Unclean Indoor Air Act of 1987," and recalls

> It no longer prohibited smoking except in designated smoking areas meeting minimum standards, but merely provided that "smoking and nonsmoking areas may be designated" in a public place. Smoking was prohibited only in a designated nonsmoking area. . . . Also stripped of the bill were all of the enforcement provisions and all of the minimum standards for protecting nonsmokers except for enforcing no smoking in the nonsmoking areas by asking smokers to refrain from smoking in nonsmoking areas upon request of a client or employee suffering discomfort from the smoke. Worst of all was the addition of the dreaded "preemption" section that prohibited cities and towns from enacting laws that are "more stringent than those of this act." As a result of the different versions passed in each house, SB327 was referred to a Conference Committee. What came out of conference was worse yet. It now mandated the designation of smoking and nonsmoking areas . . . and the dreaded "preemption" section was still intact. (Barnes 2001a)

In order to ensure that any authority to enact smoking restrictions would be confined to the state legislature, where the tobacco industry could centralize its lobbying efforts, the 1987 preemption clause prohibited local municipalities from enacting tobacco control ordinances more stringent than the clean indoor air statute. State preemption of stronger local clean indoor ordinances has never been repealed (Smoking in Public Places Act 1987).

CONTINUING INDUSTRY DOMINANCE IN THE 1990S

In 1994, House Bill 2129 amended SB 327 by subsequently allowing restaurants to establish smoke-free facilities if they chose (Barnes 2001a; McGrew 1996). Section 6 of HB 2129 mandated that the Oklahoma State Department of Health (OSDH) promulgate rules and regulations implementing the Smoking in Public Places Act (Smoking in Public Places Act 1987). By 2001, and in accordance with the law, the rules circulated by the OSDH included a requirement that state or local governments develop a written policy describing criteria for designating public smoking and non-smoking areas and preventing smoking in non-smoking areas (Oklahoma State Department of Health 2001a). Enforcement and penalties for violation of the Smoking in Public Places Act and the Department rules included the issuance of a statement of state law violation along with a possible further penalty that could include a usually unspecified and discretionary administrative penalty (Oklahoma State Department of Health 2001a). These rules and administrative enforcement requirements by the Department followed very closely the legislative intent of the very weak provisions and enforcement penalties of Oklahoma's Smoking in Public Places Act.

Despite these meager gains, the industry managed to keep significant legislation from taking place, and their use of the Oklahoma Restaurant Association to fight against smoke-free workplaces and public places would become particularly strenuous during the clean indoor air campaigns of 2002–2004 (Philip Morris 1998). In a recorded interview with Richard Barnes of the American Lung Association, one of the authors of this book remarked to him that "sounds like prior to 2002 there really wasn't much activity with regard to clean indoor air." Barnes replied, "Oh, there was plenty of activity, there was just no success." (Barnes 2004b). Reflecting the dominant role of tobacco industry lobby, the extent of clean indoor air in Oklahoma law prior to 2001 consisted of smoking prohibition—unless in designated non-smoking areas—only in: theaters, libraries, museums, art galleries, roller rinks, meetings of public bodies, health care facilities, educational facilities, licensed nursing facilities, and child care facilities. Designated smoking areas were also allowed in state and local government offices and buildings, public meeting rooms, public conveyances, museums, and concert halls (R. J. Reynolds 2001b). Restaurants were allowed to have designated smoking or non-smoking areas or be entirely designated as smoking or non-smoking. Smoking was prohibited in elevators and buses (R. J. Reynolds 2001b). Additionally, local governments were preempted from passing stronger clean indoor air legislation than the weak state clean indoor air law. Of course, the continuing problem with designated smoking areas from a public health perspective is that they do not shield nonsmokers from the impacts of secondhand tobacco smoke, which could drift to all areas of the public space (Givel and Spivak 2011–2012).

The tide slowly began to turn during the 2002 and 2003 legislative sessions. The primary catalyst for strengthening tobacco control and health advocacy with regard to clean indoor air issues was the arrival in June 2001 of Dr. Leslie Beitsch, MD, JD, formerly Deputy Secretary for the Florida Health Department, appointed as Oklahoma's new Health Commissioner (Bisbee, 2003). An outsider to Oklahoma, Beitsch was educated at Harvard and Georgetown, and had begun his career in public health in 1981 at the Broward County Health Department in Florida. Two years later he was in Tallahassee as the Division Director for Family Health Services, and was already involved in tobacco issues. Nearly two decades later, as Deputy Secretary, a colleague informed him that the Oklahoma Board of Health was searching for a new health commissioner. At first he was hesitant, but knowing that Oklahoma was one of only four states in which the commissioner reports to a board of health instead of directly to the governor, as was the case in Florida, helped encourage him to consider the job.

> We [Florida] had a fair amount of that [turmoil] with change of governorship and legislative changes. The idea of going through another transition . . . when you're working for the government and you're beyond sort of a bureau chief's level . . . you are essentially interviewing for your own job again . . . and so it didn't have that much appeal to me to do that again and quite frankly I wanted to see what I had learned and test out the experiences I had in another setting . . . the potential grew for some stability and that appealed to me after multiple changes in governorship and political sort of changes in Florida—I thought that sounded good, and a chance to make some long term impacting changes . . . and they wanted someone who was really aggressive, someone who was willing to make substantial change. And they had decided that tobacco issues would be a really important, in fact, the most important thing on their agenda, and all of that resonated with me. (Beitsch 2012a)

He also noted that in addition to the board's enthusiasm for hiring a commissioner who would take a pro-active stance on tobacco issues, Oklahoma's Governor, Frank Keating, would also be a potential ally:

> I was very impressed by the people that I met and by their commitment to improve and help Oklahoma, including the governor who was Frank Keating at that time . . . Getting him on board was part of a legacy. So when you're visiting with somebody who is in their seventh or eighth year and they aren't up for re-election again and so, 'How do you want to be remembered?' . . . [He was] someone who was supportive of health and I thought he was a much stronger ally than most people recognized or appreciated. And I was glad to work with him; I didn't get a lot of his time, but when I asked for it and told him what I wanted and why, he turned out to be a pretty supportive guy. (Beitsch 2012a)

In addition to the encouraging support of Oklahoma's Board of Health and Governor Keating, Beitsch discovered that the political landscape regarding the potential to be an advocate was quite different from what he had experienced in Florida. While health advocacy groups, such as the "tri-health" associations of American Heart Association, American Lung Association, and American Cancer Society, were extremely active in his home state of Florida, this was not the case in Oklahoma.

> Florida was also a political environment, a different political environment than in Oklahoma . . . the issues around tobacco control were led [in Florida] by the voluntary health organizations, so heart, lung, and cancer [American Heart Association, American Lung Association, American Cancer Society]. And they did an extraordinary job—they were vigorous, very organized, very outspoken and we (the Florida Health Department] were not instrumental in that process . . . it was like we were essentially handcuffed in that process. (Beitsch 2012a)

> The [Florida] Governor [Jeb Bush] . . . . made it very clear to us that there was not going to be a role for executive branch agencies in advocating for tobacco control . . . and so these same voluntary health organizations [in Oklahoma] that were sort of half in—they were kind of on the fence—you couldn't really tell if they were in it or not. In Florida, they took the lead and it was the health department that was on the fence. I mean it was really shockingly embarrassing. (Beitsch 2012b)

> The environment in Oklahoma was a hundred eighty degree difference from that. One of the reasons I went to law school ever so long ago wasn't to become a particularly skilled lawyer, because I'm not one. Because I'm about policy . . . and so for me this [was] an opportunity to shape some of those policies in a meaningful and important way and I really wanted to do that. (Beitsch 2002a)

While the political environment was more conducive toward an active health department role in tobacco control, Beitsch was still aware that Oklahoma would be a difficult environment in which to initiate tobacco policy changes.

> Tobacco is still the leading cause of death in our country and Oklahoma was much more of a quote "tobacco culture" than the average U.S. state and in fact having grown up in the south and lived in most southern states during my childhood almost exclusively I understood the notion of what tobacco culture was like. The fact that no one actually grew tobacco in Oklahoma was the only surprising element that was missing. (Beitsch 2012a)

Not only did the Board of Health want the new commissioner to be an aggressive advocate for policy, they also wanted, for the first time, an outsider who would help to counter the recent corruption scandals that had tarnished the Department of Health. The former commissioner had resigned, and one of the deputy commissioners was arrested by federal authorities in a nursing home scandal. Beitsch was acutely aware of the morale and reputation issues that the agency was suffering.

> So there's this two-year history of scandal, having the deputy health commissioner carted off in handcuffs . . . I had yet to talk to anybody that I know in public health or anyone on the planet who said "Oh, yeah, my deputy health

> commissioner was arrested by the FBI . . . that's a pretty unusual thing. And others from within the department convicted, so [there was] scandal and corruption. (Beitsch 2012a)

> [It was] a department that was completely demoralized. You had to know absolutely how deep seated the corruption was . . . people in that building would say I work in the such-and-such division, but they didn't say I work for the Oklahoma State Department of Health because they were too embarrassed. (Beitsch 2012a)

Despite the challenges, Beitsch accepted the job, and began advocating tobacco control even before arriving in May 2001. The previous month, he had sent a letter (coauthored with Oklahoma's Board of Health president, Dr. Jay Gregory) to Oklahoma's House Speaker, Larry Adair, urging the legislature to reconsider the 2001 Oklahoma Tobacco Use Prevention and Cessation Act, stating that "while undoubtedly well intended, shifting the Oklahoma anti-tobacco efforts from the Department of Health to another state agency runs counter to successful campaigns in other states" (Phillip Morris 1998). When he finally took office, he quickly discovered how unpleasant relations with the legislature could be.

> The first day, literally, I was in my office, so that would have been a Monday and just after Memorial Day, I get a call from a house member who was on the appropriations committee. He said something to me along the lines of "You know, Commissioner Beitsch, welcome aboard, I understand you're trying to get, that you all have a budget request looking for resources." . . . He said "I can help you with that but you need to do something for me." And I said, "Representative, you know today is my first day on the job and it's great to hear from you. Perhaps you haven't read a newspaper for the past two years. You know I represent a change in how business conducted here and that's not how we do it." *Click.* That was the end of that conversation. So that was sort of the first salvo for me . . . . that set a tone very early . . . and I was not happy. I mean that just infuriated me. I thought I was being tested. It bewildered me as much as anything . . . . you know, several people convicted of felonies, one hundred or there about ghost employees in the department, and I'm getting a call from a legislator asking for a quid pro quo that I couldn't possibly condone under any set of circumstances and certainly not with that track record, [that's] why I came there. (Beitsch 2012a)

Still determined to actively promote tobacco control policies as directed by the Oklahoma Board of Health, Dr. Beitsch achieved his first success just a few months after taking office in what he publicly announced was "the first phase of an 'all-out attack' on tobacco products" by banning smoking at health department main offices in Oklahoma City, health department offices in all sixty-eight counties, state vehicles used for health department business, and by health department employees providing services in clients' homes (Green 2001; Oklahoma State Department of Health 2001b).

The new policy became effective at the beginning of 2002. Beitsch announced that "we need to get our house in order first" and noted that Oklahomans who arrive at health department facilities for needed services, such as obtaining birth and death certificates, as well as pregnant women seeking maternal care, should not have to face a gauntlet of smoking at the entrance. Ken Nance, the current lobbyist for Brown & Williamson and Lorillard, responded publicly against the new policy, stating, "we have a good law as it is now" (Associated Press 2001).

The reason the new health commissioner was able to implement these new policies had to do with the political structure of Oklahoma's governorship and bureaucracy. The state's executive branch has been historically fragmented, with ten separately elected administrative offices other than the governor, including the lieutenant governor, three corporation commissioners, labor commissioner, superintendent of public instruction, state treasurer, attorney general, insurance commissioner, and state auditor (Morgan, England, and Humphreys 1991; Oklahoma Department of Libraries 2007b). The governor does not directly control any of these statewide offices (Morgan, England, and Humphreys 1991; Oklahoma Department of Libraries 2007b). Beyond these elected offices are various boards and commissions including the Oklahoma Board of Health whose members serve for nine years and oversee the Oklahoma Health Commissioner (Morgan, England, and Humphreys 1991; Oklahoma Department of Libraries 2007b). This political structure, including long terms of service by members of the Board of Health, enabled the Department of Health to engage legally in significant political advocacy (Barnes 2001b).

In tandem with banning smoking at all state and county health department offices, Beitsch approved beginning in October 2001 a voluntary and comprehensive tobacco cessation assistance program for state health department employees (Crutcher 2001). Included in cessation services were free supplies to participating employees of Bupropion medication to reduce nicotine craving, nicotine patches, nicotine gum, nicotine inhalers, and nasal sprays (Oklahoma State Department of Health 2001c, 2001b; Miner 2001). Supplies for state and county employees were provided for free for three months up to April 26, 2002. The program concluded on June 30, 2002 (Oklahoma State Department of Health 2001d; Oklahoma State Department of Health 2002).

Soon, it was clear that the new health commissioner would rigorously pursue public health policy to counter tobacco use. His "sometimes brash style and zealous pursuit of anti-tobacco policies irritated some policy-makers in Oklahoma" (Jenkins, 2003). However, former legal counsel for the American Lung Association of Oklahoma Richard Barnes recalled how Beitsch's aggressive advocacy enhanced the development of tobacco control efforts in Oklahoma.

> He came in June of 2001. He is the new kid on the block . . . he comes here from Florida. And then three months later 911 happens and then everything was diverted to bio-terrorism so it was . . . in early 2002 that is when Dr. Beitsch became very vocal and very forceful in pushing for clean indoor air. Because Florida had already taken the first step . . . he had been involved in that battle down there and he became literally like a lightning rod. And just caught all kinds of hell from the legislature and the Daily Oklahoman. (Barnes 2004b)

Doug Matheny, Chief of the Tobacco Use Prevention Service, and Bob Miner, Clean Indoor Air Program Coordinator at the Oklahoma State Department of Health, described the new commissioner in comparison to his predecessor, Dr. Jerry Nida.

> One year Dr. Nida called me into this office and said they had gotten a call from Benny Vanatta [lobbyist for the Oklahoma Restaurant Association] . . . . and they called Dr. Nida and said that I needed to back off. That the meeting needs to back off whatever this preemption bill or whatever by pushing that and Dr. Nida told Kevin and Kevin told me that yeah he knew that Dr. Nida supports what you are doing, the message that you are sending about what is important and what would be helpful and so forth. Whatever was appropriate I was doing there. But I was supportive of it but you can't do it this month because it is May and because Benny Vanatta is very influential with the appropriations committee . . . Benny Vanatta said we are going to hurt you with your appropriations and the implied message that Dr. Nida got that had been communicated to me . . . I would predict that if Dr. Beitsch had gotten a call like that that Dr. Beitsch would have said well thank you very much, I appreciate your being on that. And he would have picked up the phone and called a reporter . . . he would have seen that as an opportunity to expose Benny Vanatta. (Matheny and Miner 2004)

Nida resigned in July 2000 and later pled guilty in October 2001 to a felony charge of criminal conspiracy to defraud Oklahoma by allowing a "ghost" employee to collect a salary when he actually was not working for the state health department (Ellis 2002).

BANNING SMOKING IN THE CAPITOL BUILDING

At the same time Dr. Beitsch's ban on smoking at the health department took effect, new efforts toward clean indoor air were beginning to stir. In January

2002, state representative Ray Vaughn (R-Edmond) introduced House Bill 2401, a sweeping clean indoor air measure that would prohibit smoking in "any place that is used by the general public to which the public is invited or in which the public is permitted," and would also prohibit smoking within 25 feet of all "fresh air intake" entrances to the applicable facilities. While a coalition of 19 state health organizations voiced their support (Ford 2002b), Lorillard lobbyist Brian Nance publicly called the proposal "just a little ridiculous." (Ford 2002b). Others agreed. Rep. Michael Tyler (D-Sapulpa) expressed concern over the 25-foot rule. "I'm from a small town and we have small sidewalks. Twenty-five feet is in the middle of the street" (Associated Press 2002b). House Speaker Larry Adair (D-Stilwell) immediately assigned the bill to the Commerce, Industry and Labor Committee. Vaughn was disappointed, stating that the commerce committee was unfriendly toward tobacco restrictions, and would have preferred that it had been sent to the Health Committee. His fears turned out to be accurate, as Commerce Committee chair Rep. Lloyd Fields (D-McAlester) pronounced the bill "dead on arrival." After the pre-legislative meeting, Fields and fellow committee member M. D. Leist (D-Morris) both smoked cigarettes in Leist's office while telling a reporter that the bill would have been a "bind on business" (Ford 2002b, 2002c). A *Tulsa World* editorial derided Leist's attitude. "HB 2401 is a serious piece of legislation, intended to address a serious issue . . . but what it got was a quick kiss-off and good ol' boy benediction from Rep. M. C. Leist, D-Morris, who expressed fear that the measure might prohibit smoking at rodeos and coon hunts. Now there's a genuine contribution to the public health debate" (Ford 2002c).

In March, a Cookson, Oklahoma man, Larry L. Krantwashl filed a federal lawsuit in the U.S. District Court for the Eastern District of Oklahoma. Krantwashl, a former smoker who suffered from emphysema and chronic obstructive pulmonary disease, and had been carrying an oxygen tank since 1994, claimed that the House Commerce, Industry, and Labor Committee had denied his rights under the Americans with Disabilities Act of 1990 (ADA) with regard to public accommodations access to a smoke-free environment. In the complaint, he accused committee members of "being influenced by the tobacco lobbies to keep HB 2401 from being voted on," and that "this committee is refusing the electorate in the House of Representatives the right to enact legislation to coincide with the ADA" (Price 2002b; Ervin 2002a; Associated Press 2002d). The suit named all 16 members of the committee, each of which was served by a federal marshal on March 20, and asked for compensatory damages of $10,000 per defendant, plus punitive damages of more than $1 million and a house floor vote on HB 2401. Committee chairman Fields denied having been influenced by tobacco lobbies (despite having received $1,775 from tobacco and restaurant lobbies between 1998 and 2001); other committee members called the lawsuit frivolous and ridiculed it (Ford 2002e). Rep. Richard Philips (R-Warr Acres) produced a glass jar for fellow representatives to make contributions to his defense fund, and Rep. Russ Roach (D-Tulsa) stated "I'm going to send some old Perry Mason tapes for Drew [Attorney General Drew Edmondson] to look at so he can defend us" (Ervin 2002a).

A concurrent attempt to enact clean indoor air legislation during the Spring 2002 legislative session came from Sen. Ben Robinson (D-Muskogee), who introduced SB1553 into the state senate. This bill would originally have been very similar to Vaughn's bill, making all public places and workplaces smoke free by banning smoking in any building where the public was "invited." The measure was passed though the Senate's Human Service Committee with an amendment that exempted restaurants, but gave them a sales-tax incentive (a 3.4 cent offset on the 4.5 percent tax rebated to the state) to go smoke-free (Jenkins 2002a; Associated Press 2002c). By the time a floor substitute passed the Senate, all reference to restaurants and other public places had been removed, leaving only the State Capitol Building (but not other state buildings), and places already covered by existing law, such as movie theaters, other indoor theaters, libraries, art galleries, museums, and skating rinks smoke free (Ervin 2002a). Even this watered down bill encountered resistance when it reached the House of Representatives. House Speaker Adair assigned the bill to the Commerce, Industry, and Labor Committee, despite Ray Vaughn's (the bill's House Sponsor) request to have the bill reassigned to the Health Committee. The speaker stated that "As a nonsmoker, I am not opposed to Rep. Vaughn's ban on smoking in the Capitol building, but as speaker, I have responsibility to follow procedures and be consistent in bill assignments" (Hobcrock 2002).

The controversy heated when Judy Dobelbower, legislative assistant to Rep. Carolyn Coleman (R-Moore) since 1996, resigned in late March, citing respiratory and other health problems, including aggravation of her migraine headaches, associated with the secondhand smoke from lawmakers' offices (Talley 2002c; Ford 2002c). Coleman announced that she supported her former assistant's intention to seek a grievance with the Occupational Safety and Health Administration for "the unhealthy and hostile environment caused by this problem, along with remedies available to her through the workers compensation system, the Equal Employment Opportunity Commission and the Department of Labor." A petition signed by 88 people, including 18 legislators, was submitted to the House Chief Clerk and Administrator Larry Warden, asking for a review of the House's smoking policy, and Coleman asked Speaker Adair to address the issue by banning smoking at the Capitol and other state buildings in the State Capitol Park complex. Adair replied that legislators had been informed of the need to keep their office doors closed when smoking, but did not know what else could be done under current rules (Price 2002b). The issue of smoking in the state capitol building was something that Dr. Beitsch was also keen to remedy, as he had discovered that, "When I first came to Oklahoma you could smoke in the state capital itself and in individual offices and I remember visiting a particularly hostile-to-clean-indoor-air house member who delighted in having a conversation with me, right up in my face was blowing smoke with a cigarette that he had burning in his hand" (Beitsch 2012a).

Speaker Adair did urge Commerce, Industry and Labor Committee chairman Fields to allow SB1553 a committee hearing, while Fields mentioned that he wanted the bill to include a provision for designated smoking areas (Ford 2002c). After Fields proposed an amendment to that effect, Rep. Russ Roach (D-Tulsa) argued that such an amendment would allow legislators to continue smoking in their offices, since these could simply be designated as smoking areas. Roach then amended Fields' amendment, requiring smoking areas to be equipped with separate ventilation systems, without which, he claimed, the bill would simply provide "the illusion that we've done something" (Talley 2002e). With Roach's amendment, the bill passed through the committee, and a floor substitute expanded the measure to include all state buildings, also prohibiting smoking within 25 feet of an entrance or exit, and allowed county and municipal government buildings to enact similar restrictions. While the bill passed the House with an overwhelming 82–14 majority, some lawmakers were vocally opposed to the floor substitute. Rep. Bill Graves (R-Oklahoma City) stated, "I think we're being stampeded by what somebody has called the 'health fascists'" (Price 2002c). Nonetheless, Governor Keating signed the bill on April 17, 2002, calling it "a magnificent bill," and stated, "this is a first step, but a very important step, in eliminating secondhand smoke in areas of public access" (Ervin 2002b).

Dr. Beitsch wasn't thrilled with the provision for a smoking area, but acknowledged that the new rule for the capitol was an improvement, and he was able to implement a degree of enforcement from time to time. "What we got out of it was no more smoking within the capital and they put a smoking room in the basement that had ventilated externals, so it was a negative pressure smoking room. So every time I was in the capital I would go check to see if it was working properly. It almost never was and I would tell the engineering people to close it off" (Beitsch 2012a).

USING BUREAUCRACY TO INFLUENCE POLICY: THE EMERGENCY RULES

Despite the gains made in health department buildings, health department vehicles, clients' residences, and the state capitol, Beitsch and the board of health were still frustrated by the mediocre progress of Oklahoma's lawmaking bodies on the secondhand smoke issue. They decided to extend the health department's internal smoking ban policy to all public places and workplaces in Oklahoma via permanent amendments to the Board of Health administrative rules. The Administrative Procedures Act in Oklahoma allows state agencies, such as the health department, to enact new rules (amendments) to take effect during the legislative session that would serve as permanent state law, as long as they are signed by the governor and not blocked by a majority of both legislative houses. Emergency rules can become effective upon the governor's signature even when the legislature is not in session, but they are only effective for one year, while the legislature has the opportunity to repeal them over the following legislative session (Oklahoma Administrative Procedures Act 2004). According to Richard Barnes, plans for the emergency rules had already been in the works for a year.

> It was at the world conference on health or tobacco . . . it was in the U.S. that year [Chicago, 2000]. Doug Matheny and Bob Vincent who was then Doug's boss at the health department [Deputy Commissioner for Health Planning and Policy Development] and I were sitting in the lobby, and Dr. Vincent said Richard do you think the board of health, the state board of health using its rule making powers can do something about regulating smoking in public places . . . would you write the legal opinion to me outlining what you think the power of the board of health is to do something to rulemaking . . . . I said well don't you have lawyers in the health department and he says yes but they always tell us we can't do anything. I want someone from the outside to look at it and tell me what we can do. So I concluded they could do about anything they wanted . . . he [Dr. Beitsch] was not the thing that started it . . . it was already in the works. But him being vocal and visible in support of it which the prior commissioner had not done and probably wasn't going to do but with Beitsch being as out front as he was was [sic] a major help getting everything to happen. (Barnes 2004a)

Meanwhile, Joy Leuthard, Director of Oklahoma Alliance on Tobacco or Health and Director of Health Care Policy and Research at the Oklahoma State Medical Association, recalled the suggestion of health department rules for clean indoor air as a joint effort of non-governmental health organizations.

> One of the things that Alliance members posed to the health department was the possibility of the health department really looking at their legal purview as

> the state agency that looks at the health issues to possibly pass rules that would require clean indoor air . . . I mean, we were really working as partners. All of our organizations: Heart, Lung, Cancer, OSMA [Oklahoma State Medical Association], all of these working together to try to get that through so eventually the legislation was introduced. (Leuthard 2004)

Dr. Beitsch himself, speaking more than a decade later, said that "I don't remember where the idea came from . . . whether it was my idea or Richard's [Barnes] idea, or our lawyers' idea, I don't know, but what I remember very clearly though is that once we made a decision to go on that pathway, it made a big stratégy for us on how to proceed incrementally" (Beitsch 2012a: 8).

For a brief period in the fall of 2001, the health department's focus was redirected toward bioterrorism due to 9/11, but clean indoor air tobacco issues soon regained attention (Barnes, 2004a). Bob Miner, former TFOC leader (1997–1999) recalled the final process for implementing the emergency rules.

> The State Board of Health had tentatively planned for many months to use their administrative rulemaking authority in 2002, if the legislature had not yet acted, to extend effective protections from secondhand smoke under the Smoking in Public Places Act. It was decided to do this incrementally at their February and March meetings, with the first session promulgating rules for six types of healthcare facilities plus ambulances, all licensed by the Department of Health. In March, rules were promulgated for seven additional types of establishments licensed by the Department of Health, plus a set of more comprehensive rules covering most indoor public places and indoor workplaces with only a few exemptions. There were allowances for smoking rooms if such rooms met several requirements to effectively protect the air in nonsmoking areas from contamination by secondhand smoke. The governor has 45 days to approve such permanent rules, and his signature is required. The legislature can block administrative rules only by majority vote of both houses within 30 legislative days. (Matheny and Miner 2004b)

Prior to adoption of the rules, the health department engaged in detailed internal analyses and revisions encompassing several drafts (Nichols 2002; Slaymaker 2002; Oklahoma State Department of Health 2002). The Board of Health passed two sets of administrative rules in early 2002, one set that prohibited smoking, with some exceptions, in health care facilities and the second that covered all public places and workplaces, including restaurants. In particular, the rules provided clarification regarding responsibility for persons who operate enclosed areas to minimize or eliminate secondhand smoke health dangers (Oklahoma State Department of Health 2002). The official and scientific need provided by the state health department to Governor Frank Keating for adopting the rules related to the danger of breathing secondhand tobacco smoke included, "Secondhand tobacco smoke exposure is harmful to the health of nonsmokers. Chronic exposure can cause morbidity and mortality from heart disease, stroke, lung cancer and other cancers, and respiratory diseases. Even occasional, brief exposure can be harmful" (Oklahoma State Department of Health 2002).

Specifically, the rules covered restaurants with seating above 50, bars, taverns, pool and billiard halls, bingo halls, bowling alleys, adult day care centers, ambulatory surgical centers, birthing centers, covered malls and shopping centers, up to 75 percent of total hotel or motel rooms, youth camps, and indoor workplaces with 15 or more employees (Oklahoma State Department of Health 2002; Keating 2002a). Restaurants with seating capacity above 50, bars, taverns, and pool and billiard halls, were provided the option of being entirely smoking, entirely smoke free, or smoke free with designated smoking rooms that vented 100 percent of the smoke outdoors. Bingo halls, bowling alleys, adult day care centers, ambulatory surgical centers, birthing centers, indoor workplaces with 15 or more employees, and covered malls and shopping centers were to become smoke free unless they provided designated smoking areas. Hotels and motels were given the option of designating up to 75 percent of rooms smoke free with hotel and motel lobbies being smoke free. Youth camps were totally smoke free.

Richard Barnes described the political atmosphere and media attention associated with waiting for the gubernatorial response, even with their various exceptions that allowed smoking in designated areas or in some cases allowed smoking entirely.

> So we started the pressure on him on the governor to sign the darn things. I think it was April Fools day. It was the first day that the ads [in newspapers] started running . . . which I believe was a Monday. By Thursday the governor's chief of staff called the lung association because we were the only organization named as the sponsor on the ads . . . and said can you please do something to pull those ads. You have shut down our switchboard. They were getting that many calls. (Barnes 2004b)

Meanwhile the ORA submitted a letter to the Oklahoma State Department of Health raising several concerns. ORA's primary concerns and objections included questioning whether the rules were constitutional and whether the rules were "vague and unclear." Further, the ORA expressed the concern that "our members should not be put in the position of being a policeman to their customers" (Oklahoma Restaurant Association 2002a).

Governor Keating, who had stated that he was "viscerally in favor of a smoking ban in restaurants" (Talley 2002a), signed most of the first but none of the second, concerned with the legal implications of a state agency making new law that conflicted with current law, specifically the 1994 amendments to the 1987 Smoking in Public Places Act. In an April 5, 2002 letter to Commissioner Beitsch, the governor regretfully stated his objections.

> I wholeheartedly support the Oklahoma Board of Health's efforts to reduce the incidents of smoking and the negative effects of secondhand smoke, within the confines of the law. However, I must disapprove all rules in the second set regarding smoking in public places other than health care facilities. A great majority of these proposed rules directly conflict with the Smoking in Public Places Act . . . when the law specifically authorizes smoking in a public area, the Oklahoma Board of Health cannot pass a rule prohibiting it . . . again, I do support your efforts, but believe we must respect constitutional principles such as the doctrine of separation of powers. While I must disapprove these rules for the legal reasons stated above, I am calling on the Legislature to change the law to prohibit smoking in most public places. Their action would in effect implement your rules in a legal fashion. (Keating 2002b)

That same day, the governor's office issued a press release in which Keating stated that he agreed with "the sentiment" of the rules, and that "the answer is for the legislature to take a bold step for public health and rewrite the law . . . if no action is taken this session, I will work with the Health Department to develop strong rules which restrict smoking but conform with the law" (Keating, 2002a). When the governor signed SB 1553 a few weeks later in a ceremony in the Blue Room at the Oklahoma State Capitol, he repeated his challenge to the legislature to change the existing statutes, and promised "to sit down with the Health Department and help its officers write rules 'that will pass legal muster' to ban smoking in public places such as restaurants if the Legislature fails to do that in the six weeks remaining before adjournment" (Ervin, 2002b). Afterward, his staff privately asked some of the public health group leaders to tone down their campaigns. All the phone lines to the governor's office were being tied up with constituents calling to support both the rules as well as new legislation on clean indoor air (Matheny and Miner 2004b).

The legislature indeed failed to process any legislative initiatives, spurring the Health Department, its legal representation, the TFOC leaders, and the governor's staff into a frenzy of drafting new rules. Their deadline was the June 6, 2002 Board of Health meeting. According to the Administrative Procedures Act in Oklahoma, emergency rules can be enacted outside of the legislative session with the governor's signature. Though the rules would only be in effect for either one year or until the legislature reviewed them, this Act offered a chance for the TFOC and its allies to make a strategic move. Matheny and Miner recalled the process of working with the governor's staff in the following months.

> This set of rules was another Chapter 99, to cover a broad array of indoor public places and indoor workplaces, with only a few exemptions. A fresh approach was taken, this time taking pains not only to be consistent with existing statutes, but also to overcome the objections the governor had had with the rules promulgated in March. A very strong set of rules was developed and approved unanimously at the Board of Health meeting on June 6. These were in many sections so that small individual components could be disapproved without disrupting the entire package. The governor could pick and choose to some extent. We were given assurances by the governor's staff that he would probably sign the public places sections; they were not at all sure about the private workplaces rules, but they were left in the package in hopes that they too would be approved. The public health groups continued their show of support, encouraging gubernatorial approval of all of the new emergency rules. On June 27, with only three hours notice, Governor Keating surprised almost everyone with a news conference to announce he was signing all of the clean indoor air rules adopted by the Board of Health and that they would be effective July 1. (Matheny and Miner 2004b)

Many of Oklahoma's legislators felt that the governor and the Health Department had overstepped their constitutional authority in this regard, and that unelected officials did not have the right to make policy. The tobacco lobby and its allies certainly felt this way. In March, Benny Vanatta, lobbyist for the Oklahoma Restaurant Association, stated "We think the commission has way overstepped its bounds and is making new law" (Price, 2002d), and followed in July with the comment that "I hope Dr. Beitsch is a better doctor than he is an attorney" (Barnes 2004a; Talley 2002b). Regardless of the legality of the rules, Richard Barnes credited the controversial publicity with helping to pass SB 1553, as well as inciting future legislation initiatives.

> What those Board of Health rules did and this is crucial to the whole story . . . they passed the first set of them like the middle of February . . . and of course it hit the news . . . and the news coverage, the editorials, the letters to the editor on the pros and cons of regulating smoking in public places was just enormous and we already had our battle plans laid for editorial boards, board meetings and all of that. And that is the only reason the bill [SB1553] passed that banned the smoking in all state owned and operated buildings in counties and cities to be smoke-free is because of the publicity that those board of health rules generated . . . Which was just enormous. I mean we were getting two and three stories a week about what was going on. You couldn't pay for this kind of stuff . . . it was all just earned media and then in early June after the session is over and we worked in April and May with the governor's staff to come up with rules that the governor would sign. And the board of health passed them in June. The governor signs them and we have a big press conference about it and the governor is telling them—I told the legislature if they didn't take care of it I was going to do this. They didn't take care of it so here we are doing it. (Barnes 2004b)

FIGHTING THE LEGISLATURE AND THE TOBACCO INDUSTRY

Dr. Beitsch was immediately subject to backlash and criticism from both legislators as well as *Daily Oklahoman* newspaper editorials. He recalled that in Oklahoma

> The legislature was by far the strongest of the branches of government . . . not only were they opposed to changing the status quo, it was also this issue of "we'll make legislation when we get around to it . . . . and butt out and who told you *you* had any rights?" and so the idea that the board of health and the health department wanted to take an active role in policy making, that was usurping their prerogative and was not going to be well received by people who viewed themselves as very much entrenched. (Beitsch 2012b)

Simultaneously, the *Daily Oklahoman* ran nearly a dozen editorials, as well as about two-dozen letters to the editor, between the summer of 2002 and spring of 2003, about the clean indoor air rules. Most of the opinions expressed were negative, deriding Beitsch and the health department for being overzealous and tyrannical. Titles included "Snuffing Out Choice" (2002e), "Dr. Beitsch, Health Thyself: Commissioner His Own Worst Enemy" (2002c), "Above the Law: Zealots Taking Over Policy Process" (2002a), "Cooking Up Confusion: Beitsch Rules Blindside Small Business" (2002b), "The Shadow Legislature" (2002d), and "Phony, Hypocritical" (Sieber 2002).

The controversy quickly manifested into a legal battle. On June 27, 2002, the day after the governor signed the new emergency rules, Sapulpa plaintiffs Freddie's Barbecue and Steakhouse and the Veterans of Foreign Wars Post 1320, in consultation with the Oklahoma Restaurant Association, filed suit with the Creek County District Court (Bellamy 2002c). The suit, which named the Oklahoma State Department of Health and Governor Frank Keating, claimed that the rules were more stringent than state law and ran contrary to the intent of the legislature. As the restaurant association was organized and funded by the tobacco industry, the health department immediately understood what kind of opposition they faced (Spivak and Givel, 2005; Kumpe 2002; Williams 2002). Dr. Beitsch recalled that "the restaurant association is a stomping place for the tobacco industry . . . and so when the restaurant association is challenging you, you know who's really challenging you" (Beitsch 2012a).

Keating and the Health Department attempted to have the lawsuit moved to the U.S. District Court in Tulsa; health department attorney Charles Broadway claimed that "there are a number of constitutional issues that have been brought up, and we feel that the federal court is the best arbitrator for those

. . . the Americans With Disabilities Act also may come into play because many people have ailments that preclude them from being around cigarette smoke but who still want to eat in restaurants" (Barber 2002). On July 12th, 2002, U.S. District Court Judge James H. Payne ruled to send the suit back to state court in Creek County, stating that "the controversy will be resolved by a determination of whether defendants have exceeded their authority under Oklahoma statutes, and whether the rules at issue conflict with Oklahoma law . . . no substantial question of federal law must be answered to determine plaintiff's claims" (Bellamy 2002a, 2002b).

The Creek County judge, Donald Thompson, was distinguished in two ways. First, he had officiated at restaurant association lobbyist Benny Vanatta's wedding (Beitsch 2012a), which the defendants believed was a conflict of interest. Second, he would a few years later be convicted of indecent exposure in association with the use of a sexual device during court proceedings, and sentenced to a four-year jail term (Associated Press 2006). The defendants attempted to get Judge Thompson to recuse himself on the basis of his association with Vanatta, but he refused (Beitsch 2012b: 4). Further troubling was the fact that the plaintiff's attorney, Mike Morgan (D-Stillwater), was an influential member of the legislature (Beitsch 2012b; Bellamy 2002b).

> Our senate appropriations chair, not just the Health Care Committee, the entire appropriations committee, was the opposing attorney in this case and it puts you in an awkward position as you might imagine. One you still want to win so you don't want to embarrass the opposing side too much, because they have a lot of ways to even out the score. (Beitsch 2012a)

> So, I went to [Oklahoma Attorney General] Drew Edmondson and I asked him that question, "How can this be possible . . . I mean, that's got to violate every ethics law that there is." And so I pause at that point and I just get thirty seconds of laughter from Drew Edmondson, and I'm like "This is not funny to me, what's so funny?" and he said to me "What ethics laws . . . there are no such ethics rules in Oklahoma, good luck." (Beitsch 2012a)

With no remedy for ethical violations, the case proceeded in Clark County. Judge Thompson issued a temporary injunction on July 23, 2002 blocking enforcement of the emergency rules on all affected business (his previous temporary restraining order applied only to the two plaintiffs), stating in his ruling that "there is a strong likelihood that the plaintiffs will prevail on the merits" of their case (Talley 2002c; Harper 2002). The defendants were especially perplexed at the judge's ruling, and questioned the source of his written opinion.

> One of the requirements of the restraining order was that the plaintiff is experiencing severe economic damages . . . . the restaurant owner was a guy named

> Tex Schlim . . . the question we asked Tex when he was on the stand was "How has this affected your business," and Tex's response was essentially "It's never been better." At that point we don't have to say anything else . . . and so we rested our case shortly thereafter... within hours, a 95 page decision comes down from the judge, which I'm sure he had time to write. (Beitsch 2012b)

Benny Vanatta declared a victory, stating, "It's exactly what we wanted to happen. . . . It was never a smoking issue. . . . It was an issue about who had the authority to do it. They did not have the authority to make new law" (Harper 2002). Echoing the sentiment of the ORA lobbyist, Morgan told the Associated Press that enforcement of the new health department rules "would create a 'bureaucratic dictatorship'" (Harper 2002), and that "this is government bureaucracy out of control . . . this practice has to be stopped now before it goes any further" (Bisbee 2002). He called Judge Thompson's injunction "a big victory for Oklahoma businesses" (Bellamy 2002a, 2002c). The health department filed an appeal with the Oklahoma Supreme Court on August 20, 2002 asking the court to overturn the injunction and transfer jurisdiction to Oklahoma County, claiming that under jurisdictional law the venue for legal action against public officials is their county of residence. After hearing arguments in late August, the Supreme Court on September 12, 2002 unanimously rejected the agency's request for a change of jurisdiction, and the case remained in Creek County (Jenkins 2002b). Dr. Beitsch recalled, "The [Oklahoma] Supreme Court appointed a special master who was at least a hundred years old and he slept through the entire case, then issued a decision adverse to us. I couldn't believe it, I mean it was astonishing" (Beitsch 2012a).

Meanwhile, the American Lung Association of Oklahoma filed a federal lawsuit on July 18, 2002 in Tulsa, claiming that the 1987 Oklahoma Smoking in Public Places Act, which allows business to be all-smoking if they chose, violated the federal Americans With Disabilities Act. Named as defendants in the suit were the VFW post and restaurant named as plaintiffs in the Creek County lawsuit, as well as the Oklahoma Restaurant Association and the health department (the suit also challenged certain provisions of the new emergency rules). When U.S. District Judge Sven Erik Holmes denied the American Lung Association's motion for a summary declaratory judgment regarding the ADA challenge and delayed the challenge to the administrative rules pending the outcome of the Creek County case, the suit's defense attorney Sen. Michael Morgan (D-Stillwater), also the plaintiff's attorney in Creek County, said that he was "very pleased" with the ruling and that he believed that the federal lawsuit was "pretty well dead in the water" (Harper 2002). While the suit was not successful, it flushed out one crucial aspect of the ORA's political agenda. In the Response to Plaintiff's Statement of Undisputed Material Facts, the ORA stated in a footnote that, "In view of plaintiffs' desire to use this action to establish statewide smoking policy, the ORA requested, and hereby acknowledges, that Brown & Williamson Tobacco Corp., Lorillard Tobacco Company, and R. J. Reynolds Tobacco Company provide financial and legal assistance in defending this action" (United State District Court for the Northern District of Oklahoma, 2002).

Nonetheless, while the American Lung Association's lawsuit did not survive, neither did the Creek County lawsuit; the controversy over the rules became a moot issue with the introduction and passage of Senate Joint Resolution (SJR) 21 in early 2003. The development of this new legislation, which would provide for smoke-free workplaces and public places, began when the ORA announced in a press conference on December 11, 2002 that it would support smoke-free public places and workplaces legislation in the Spring 2003 legislative session (Francis-Smith 2002; Oklahoma Restaurant Association 2002b). Matheny and Miner recalled the suspiciousness surrounding the ORA's sudden turnaround.

> By December 2002, there had been reports that moderate and pragmatic forces within the Restaurant Association might prevail and lead the association in the direction of accepting some type of smoke free restaurants legislation, particularly if it would create a level playing field, that is, if it would affect everyone similarly without special exemptions for certain types of establishments. The association attracted considerable attention when they called a news conference to announce their support for a clean indoor air bill. But instead of a realistic proposal, they called for a stringent smoking ban to include bars and that it be submitted to a vote of the people, with a two-thirds supermajority required for passage. The Restaurant Association claimed to be turning over a new leaf, but the proposal seemed designated to frighten some potential supporters and to have little chance of getting a majority, much less two-thirds, of the electorate to support it if the legislature were to place the matter on the ballot. (Matheny and Miner 2004a)

Dr. Beitsch agreed that the supposed change of heart was a facade. The measure would prohibit smoking even in bars, and would be even more stringent than smoking bans in California and Massachusetts. Further, the proposed two-thirds referendum requirement that the ORA supported was contrary to the state's constitution. Matheny and Miner describe Beitsch's characteristic brazenness in reacting to the ORA's December 11, 2002 press conference.

> He was all the time doing something like that . . . the restaurant association came out and announced that they were going to support clean indoor air . . . they announced this new conference and they got wind of it in the morning and Dr. Beitsch to schedule that day so he could also be at the news conference.

> And of course it was important to his news conference he didn't try to take over . . . but all the press knew who he was and here is Dr. Beitsch the commissioner of health sitting at the back of the room so as soon as the restaurant association had said whatever they wanted to say the press was eager to gather around Dr. Beitsch. (Matheny and Miner 2004a)

Afterward, Beitsch gathered a crowd of media in the parking lot afterward and openly denounced the disingenuous tactics of the ORA, and of course their tobacco industry backers. "It is clearly the intention [of the ORA] to put on the ballot a measure destined to fail," Beitsch told reporters in the parking lot outside the ORA's December 11, 2002 press conference (Francis-Smith 2002). Even governor-elect Brad Henry remarked, "I think it's an unrealistic proposal and they probably realize that . . . I just can't imagine why in the world we would submit any vote to the people and then require a two-thirds vote for passage. It makes no sense to me" (figure 4.1) (English 2002). Benny Vanatta stated that he hoped "the results of the election would serve as an indicator to the Legislature and to future legislatures of what the people want," and when asked if the ORA was deliberately proposing a bill designed to fail, he replied, "We think it has a very good chance to pass" (Francis-Smith 2002).

Figure 4.1. David Simpson—12/18/2002. *Tulsa World*. In Bubble: "Ban Smoking in All Public Places!! Fine Print. But require a vote of the people with a two-thirds majority." On jacket of man in pocket: "Oklahoma Restaurant Association." Reprinted with permission (not an endorsement).

Meanwhile, relations between the health commissioner and the legislature were deteriorating even further. As Beitsch describes:

> [The] relationship with the house in particular, not so much the Senate, was extremely uneasy. There was a lot of friction there that thought was unnecessary. I am sure they thought I was overly heavy handed. I thought often they had forgotten who they worked for and why. And there was a lot of antagonism in that relationship, they probably felt like I didn't give them the respect they deserve and I felt they took every opportunity to take a swipe at us when they could. (Beitsch 2012a)

> I definitely received threats from the people in the legislature . . . so much so that I wasn't sure it was a good idea to have my number in the phone book, which it was, and I wasn't sure exactly how far these people would go because they seemed not be willing to control themselves. . . . I don't think many people know that—that I was sometimes concerned about the safety of my family . . . it was a reasonable concern, let me make that clear . . . there were times I wondered how far would these people go to get what they wanted. (Beitsch 2012a)

The hostility that Dr. Beitsch and the health department experienced throughout the summer and fall of 2002 took a toll, and the health commissioner eventually resigned and returned to Florida in early 2003. He recalled feeling ironic when people would ask him about his three-year experience. I'd hear: "'How'd you like Oklahoma. People are so friendly there.' It's like; you haven't met a tobacco lobbyist or legislator, have you? Even if they hate your guts, they might not tell you that right away, the first time you met them. In Oklahoma, I frequently was told in no uncertain terms that they hated my guts and I should go back where I came from" (Beitsch 2012a).

Nonetheless, the controversy over clean indoor air, initiated by his pursuit of the emergency rules, prompted the development of Senate Joint Resolution (SJR) 21. Amid a flurry of public smoking bills, only SJR 21 survived the 2003 legislative session (Bisbee 2003). The bill brought a clean indoor air standard to most Oklahoma public places and most workplaces, with some exceptions. Restaurants were given an extension for phase-in of the bill until March 1, 2006, ostensibly so that those who chose could build smoking rooms in which food and beverage may be served, but "which shall be in a location which is fairly enclosed, directly exhausted to the outside, under negative air pressure so smoke cannot escape when a door is opened, and no air is re-circulated to non-smoking areas of the building" (Miner 2004; Oklahoma State Senate 2003a, 2003b). Ventilation of tobacco smoke in separate smoking rooms, however, has been found to be an ineffective approach due to leakage in controlling the harmful health effects of secondhand tobacco smoke (Lee et al. 2010).

Otherwise, the act became effective September 1, 2003. The new law was much stronger in countering secondhand tobacco smoke than the Oklahoma State Department of Health's rule that was effective January 1, 2002. Some specified places, such as stand-alone bars, licensed bingo rooms, tobacco stores, and workplaces "where only the owner or operator of the workplace, or the immediate family of the owner or operator, performs any work in the workplace, and the workplace has only incidental public access" were exempt (Oklahoma State Senate 2003a).

SJR 21 brought the 100 percent clean-indoor-air standard to most Oklahoma public places and most workplaces, with some exceptions. Restaurants were given an extension for phase-in of the bill until March 1, 2006, ostensibly so that those who chose could build smoking rooms in which food and beverage may be served, but "which shall be in a location which is fairly enclosed, directly exhausted to the outside, under negative air pressure so smoke cannot escape when a door is opened, and no air is recirculated to non-smoking areas of the building" (Oklahoma State Senate 2003a; Matheny 2004). Otherwise, the act became effective September 1, 2003. Some specified places, such as stand-alone bars, licensed bingo rooms, tobacco stores, and workplaces "where only the owner or operator of the workplace, or the immediate family of the owner or operator, performs any work in the workplace, and the workplace has only incidental public access" were exempt (Oklahoma State Senate 2003a).

THE ROLE OF THE HEALTH ORGANIZATIONS

Prior to 2001, the health groups had primarily engaged in mostly ineffective insider lobbying tactics in the halls of power including the state legislature, administrative agencies, and governor's office. Echoing and in sync with Dr. Beitsch's public outsider campaign, a number of health organizations began to use public outsider tactics to pressure the governor and state legislature to pass clean indoor air rules and legislation. In early 2002, during the effort to convince Governor Keating to sign all of the emergency Oklahoma Department of Health rules related to smoking in public places:

> The governor repeated this call to action by the legislature at a signing ceremony for SB 1553, the measure making state buildings smoke free. After that even his staff privately asked leaders of some of the public health groups to tone down their campaign, as all phone lines to the governor's office had been tied up with calls from constituents calling first for the rules, then for clean indoor air legislation. (Matheny and Miner 2004b)

Nevertheless, the creation and impetus for the adoption of the emergency rules was due to the diligent work of a select group of OSDH employees with health advocates assisting the effort. In June 2002, Dr. Beitsch provided a public overview at the Oklahoma State Board of Health meeting regarding this effort.

> I'd also like to thank Richard Barnes who was and is the President of the Tobacco Free Oklahoma Coalition, a name of which organization, I never get right, but likewise Richard has been everywhere we've wanted him to be, except New York once, and I wished he wasn't there. The amount of time and effort and energy that Richard has put in has been beyond the expectation that anybody could have for a "volunteer." Mr. Barnes interjected, "Glad to do it."
>
> Dr. Beitsch continued, "Thank you, Richard. But the work of this rule has been a core group of the Department who worked real hard on this for literally several weeks. Somewhat of a moving and evolving time." (Oklahoma State Board of Health 2002)

Also, in conjunction with Dr. Beitsch's vigorous outsider advocacy for effective clean indoor air legislation, the TFOC engaged in a robust insider lobbying effort (Barnes 2003a, 2003b, 2003c, 2003d; Leuthard 2003a, 2003b, 2003c, 2003d, 2003e, 2003f). Organizations leading the ongoing advocacy effort to persuade state legislators and the governor to pass the legislation were the American Lung Association and Oklahoma State Medical Association. Insider tactics included one-to-one lobbying of legislators and the governor, letter campaigns, and legal analyses.

In the aftermath of the Creek County lawsuit, Dr. Beitsch and the Oklahoma State Department of Health issued a press release with respect to the tobacco industry's quiet support of the ORA's lawsuit against the Oklahoma State Department of Health (Miner 2002; Kumpe 2002; Williams 2002). However the American Heart Association with respect to the necessity, conflicting priorities, and tone of the state health department's press release advanced opposition to the press release. In particular, as the American Heart Association's Barbara Kumpe wrote on October 10, 2002:

> While I know the Health Department would like to have a scathing press release about this issue printed, if it is going out under the name of the AHA (American Heart Association), we will write our own document. As you should be aware we are supported by many restaurants across the state. While most of them maybe smoke free we cannot jeopardize our relationship with them over this issue if they are members of the restaurant association. (Kumpe 2002)

In essence, the American Heart Association of Oklahoma did not want to alienate their supporters who owned restaurants. In this case, this situation created a conflict between vigorous and highly public advocacy and advocacy to counter secondhand tobacco smoke and not unduly alienating a vested corporate interest.

Dr. Beitsch's recollection of the role of the health organizations suggests that they were not committed to the same extent as the health department.

> In Oklahoma none of these groups—certainly not all and I think how hard people like Richard Barnes worked—the pensiveness of those advocacy groups disappointed me in the fervor with which they were trying to proceed . . . I wasn't sure if the advocacy groups were "A" on our side, "B" could be trusted, and "C" would follow our direction. And that's the difference between Florida and Oklahoma. (Beitsch 2012a)

> The Department of Health was "all in" and there were times when some of the voluntary health organizations were not "all in." They were thinking about the relationships and other things that they had with funders, with legislators and others . . . they were not willing to make at times the kind of total commitment that the health department was, even though the health department had many and more downside risks than I thought they did. (Beitsch 2012b)

Although several health organizations, including the American Lung Association and Oklahoma State Medical Association, later provided insider and outsider advocacy support for secondhand tobacco smoke legislation in relation to vigorous secondhand tobacco smoke advocacy conducted by the Oklahoma State Department of Health, other health organizations did not. In addition to insider lobbying of the legislature and governor, in early 2002, the TFOC, using outsider tactics, held a press conference at the Mazzio's Corporation to announce Mazzio's support for smoke-free restaurants (US Tobacco Weekly 2002). Despite this effort, Dr. Beitsch also believed the American Cancer Society was creating the appearance of "cracks" in the TFOC (Rice 2003).

Clearly, some of the health organizations engaged in supporting outsider advocacy and insider lobbying with respect to the Oklahoma State Department of Health's outsider advocacy effort to hold politicians publicly accountable for their pro-tobacco actions. Even so, some of the health organizations were not comfortable, or at times, not supportive of the blunt outsider advocacy efforts. Instead, their orientation was geared toward quieter and less vigorous insider tactics; the latter holding politicians and vested interests publicly accountable for their actions.

SUMMARY

In summary, efforts to enact vigorous clean indoor air smoking restrictions in indoor public places and workplaces were unsuccessful during the late 1980s, resulting in a pro-tobacco industry act in 1987 that included local preemption and actually required smoking sections in restaurants. Real progress in clean indoor air did not arrive until 2003, when a Senate Joint Resolution, created in reaction to an aggressive health commissioner's public campaigning, brought about smoke-free public places and workplaces. This effort occurred due to a vigorous outsider campaign with support from Oklahoma health organizations by Oklahoma Health Commissioner Dr. Leslie Beitsch, a strategy that held politicians publicly accountable for their pro-tobacco actions. This effort reflects public administration ethics that serve the public good and promote compassion, benevolence, fairness, and optimism to resolve the injurious effects of tobacco use through effective legislation (Givel and Spivak Winter 2011–2012). The tobacco industry resisted efforts to bring about clean indoor air smoking restrictions, but despite engaging in rigorous opposition and successfully maintaining preemption throughout the state, was unable to thwart the long-fought and finally triumphant resolution that brought Oklahoma clean indoor air in 2003.

Chapter Five

The 2004 Battle for Cigarette Taxes

Like clean indoor air legislation prior to 2001, the tobacco industry was very successful in stopping the passage of any tobacco excise tax increase in the Oklahoma legislature. The momentum caused by the adoption of significantly stronger clean indoor air rules and legislation provided the impetus in 2004 for the first significant tobacco tax increase since the early 1980s. Tobacco tax increases of 10 percent per pack have been long recognized in numerous peer reviewed publications to decrease smoking prevalence by up to 5 percent (Lum, Barnes, and Glantz 2009; Chaloupka et al. 2002).

OKLAHOMA TOBACCO TAX POLICY BEFORE 2001

Oklahoma Attorney General Drew Edmondson noted in July 2004 that "Historically Oklahoma has been a low tax state in terms of tobacco products. I don't know right now where we rank among the fifty states but I would guess that we are in the bottom ten. And that is largely a result of the tobacco lobby and its efforts here in the State of Oklahoma" (Edmondson 2004). The Attorney General actually guessed quite closely. While Oklahomans voted and passed in November 2004 a referendum known as State Question 713 for a net 55 cent per pack tobacco tax increase, the Oklahoma legislature had not otherwise raised its tax since 1987, when it increased from 18 to 23 cents per pack. By 2004, before passage of the referendum, Oklahoma ranked 42nd among the 50 states for the size of its cigarette excise tax (Carter 2004c; Oklahoma Alliance on Health or Tobacco 2004c).

As early as 1983, the industry was active in fighting tobacco tax legislation in Oklahoma. In March of that year, Lorillard Tobacco Company sent notices to Tobacco Action Network volunteers in Oklahoma, informing them that the

Oklahoma Senate was considering House Bill 1325, which would increase the excise tax by four cents. The letter informed its volunteers that:

> It is our feeling that such legislation is not only unfair and unwise, but detrimental to the entire industry. If you agree, please write to or telephone your state legislators, express your opposition to House Bill 1325, and encourage them to vote "NO." The name, address and telephone number of your state senator appears at the bottom of this letter. Enclosed you will find a list of points that can be made (in your own words) in opposing this tax increase. . . . It is important that you contact your state legislators as soon as possible to ensure that your voice will be heard before final action is taken on the bill. (Lorillard Tobacco 1983)

A ten-page "Regional Excise Tax Fact Sheet" by the Tobacco Institute in February 1988 was sent by the manager of legislative issues, Paula Duhaime, to a regional staff member with a request to have a copy made for Ken Nance (Tobacco Institute 1988b). The fact sheet stated that "Oklahoma's cigarette sales, which had declined at a steady rate of nearly 3 percent annually for the previous five years, slipped four percent in 1987 . . . since then, tax-paid sales have dropped nearly 12 percent for the six months for which data are on hand" (Tobacco Institute 1988a). After noting these alarming facts about declining sales, the fact sheet served as a resource for arguments against future tax increases, claiming that further increases would cause reduced tax revenues by further decreased sales, rising illicit sales, and lower revenues for grocers and other retailers due to decreased consumption. The sheet also argued that the tax would disproportionately affect low-income households, stating that:

> More than 250,000 or 20 percent of Oklahoma households have an effective buying income of less than $10,000 per year. It is the smokers in these families who will suffer the most from an increase in the cigarette tax rate. . . . Under the current tax a household in Oklahoma with two average smokers pays $482 in state and federal taxes on cigarettes a year for the pleasure of smoking. If another 4-cent tax were to be added to the price of cigarettes, this same family would pay $526 in taxes for its cigarettes. (Tobacco Institute 1988b)

The Tobacco Institute fact sheet also purported to be concerned about the effect of a tax increase on Oklahoma's elderly, as "13.2 percent Oklahoma residents are aged 65 or over. For those elderly persons, many of whom are living on a fixed income, any increase in the state cigarette tax rate may take away this affordable pleasure" (R. J. Reynolds 1988b). Some of these arguments may have been effective, as the industry rejoiced when efforts to raise the cigarette tax failed in the 1988 session. A Choice Action Alert sent out by R. J. Reynolds in November of that year declared:

> A Victory for Smokers in Oklahoma!—Governor Henry Bellmon tried to increase cigarette taxes, but failed, at least for the time being. State officials decided against raising cigarette taxes to finance prison reconstruction. However a cigarette tax increase is likely to come up again when the legislature reconvenes in January. So write to your representatives in Oklahoma City now. The addresses at the state capitol are . . . (R. J. Reynolds 1988b)

The Institute's public relations efforts seemed to be effective. In July of the following year, R. J. Reynolds was proud to declare that:

> Smokers in Oklahoma should give themselves a pat on the back. Their phone calls and letters helped keep a cigarette tax increase off the Legislature's agenda . . . Gov. Bellmon wanted to raise your cigarette taxes by 60 cents a carton to finance a bond issue. The governor went to the Legislature with the proposal but it received little support. Legislators had received a loud and clear message from smokers that the governor's proposal was extremely unfair. (R. J. Reynolds 1989)

The industry's sophistication and influence during this early period is illustrated by the State Tax Plan composed by the Tobacco Institute in September 1989, just a couple of months after R. J. Reynolds's latest congratulatory victory statement. The heading "Objective" is followed by the statement, "To convince legislators that higher tobacco taxes should not be a part of any plan to improve the state's finances, and that the Governor's bond proposals should not be funded by tobacco tax hikes" (Tobacco Institute 1990b). To achieve this goal, the heading "General Strategy" is explicit.

> The tobacco industry has an outstanding lobbying corps in Oklahoma, led by TI [Tobacco Institute] legislative counsel Ken Nance. This group will continue its practice of gleaning advance information as to the exact nature of the Governor's tobacco tax proposals, and persuading key legislative leaders that the plans are unsound. The new House Speaker, Representative Steve Lewis who ousted Speaker Jim Barker in a power move last year, will play an important role in the resolution of the cigarette tax issue. As the Governor's precise intentions are determined, Lewis will be urged to delay referral of the bill to committee, and later refer it to a committee, which will view it with a very critical eye. (Tobacco Institute 1990b)

The heading "Resource Needs" describes the process of using "a large number of grassroots contacts" with legislators, with most of the mail generated from tax activists and phone bank operations using member-company mailing lists. Resources included a possible "media tour with TI [Tobacco Institute] economists . . . during the 1990 legislative session" (Tobacco Institute 1990b). In September 1990, R. J. Reynolds' "State Tax Plan" summarized "Industry

Action" and again illustrated the planning, organization, and resources behind industry efforts to keep cigarette taxes from being passed.

> TI [Tobacco Institute] legislative consultant Ken Nance, in cooperative effort with other industry lobbyists, is working intensely to obtain advance information concerning anticipated legislation seeking to increase tobacco taxes . . . support will be sought from organized labor members of the Oklahoma Association of Tobacco and Candy Distributors, as well as other industries subject to excise taxes and targets of tax increase proposals . . . targeted legislative leadership and executive department personnel will receive telephone calls, letters and mailgrams from TAX activists as well as interested registered voters. (R. J. Reynolds 1990)

In 1994, Tobacco Institute proudly declared that due to their lobbying efforts, "for the past seven years, under both Republican and Democratic governors, proposed tobacco tax increases have been repeatedly defeated" (Tobacco Institute 1994). The tobacco tax issue would remain dead for almost a decade. Bob Miner recalled the difficulty of supporting tax increases during the late 1990s.

> The tobacco excise tax increase was in the policy arsenal of the Tobacco-Free Oklahoma Coalition, even before adoption of a state plan with policy recommendations in January 1999. In our successful 1996 application to The Robert Wood Johnson Foundation for a three-year Smokeless States grant, TFOC stated that increasing the excise tax was a policy priority. When the RWJ (Robert Wood Johnson) selection committee visited Oklahoma I was responsible for the presentation of our support for increasing the excise tax, and I recall taking the issue to the American Heart Association, Oklahoma Affiliate Board of Directors for their specific authorization before agreeing to make that presentation. Increasing tobacco excise taxes was already a policy of the national AHA organization, but it was important to make the Oklahoma Affiliate Board aware of this fact and to assure their support. The Oklahoma AHA Board had previously voiced caution over participation in a coalition that might speak out loudly—and independently—on public policy issues. Though tax increases are especially sensitive issues, the Oklahoma Affiliate Board signed on once they understood the issue. When we made the presentation assuring the RWJ panel of TFOC's dedication to a significant increase in excise taxes, two legislators who were present also spoke in support of this as a long-term goal, but they stated it would be very difficult to get such as measure through the legislature in the near future, especially in an election year. (Miner 2004)

Efforts in the legislature to raise the excise tax on cigarettes in 1998 prompted a barrage of public relations lobbying by the industry. R. J. Reynolds instituted a campaign of cold-calling Oklahomans under the front group "Oklahoma Smokers' Rights," in which favorably responding recipients were connected directly to their state senators' offices (R. J. Reynolds 1998).

In early 1999, Governor Keating's chief of staff announced that the governor's office was considering an 11 to 13 cent increase in the cigarette excise tax in order to fund a higher education bond program (Jenkins 1999a). Talks between the Republican governor's office and Democratic legislative leaders regarding the use of a tax hike to fund the bonds drew criticism from some GOP leaders. Rep. Fred Morgan (R-Oklahoma City), House Minority Leader announced, "I'm personally not in favor of imposing a cigarette tax to pay for this" (Jenkins 1999b). The industry also responded, with Philip Morris implementing a public relations campaign by calling voters and encouraging them to call their legislators if they were opposed to the tax, providing phone numbers to the appropriate offices. An instruction in the callers' script instructs them to "give numbers to constituents who understand the issue, are articulate, and polite, and who adamantly oppose any new cigarette excise tax increase" (Philip Morris 1999b). By June, the House voted 68–18 against a shell bill that would have allowed the cigarette tax increase to go to a statewide vote. House Speaker Lloyd Benson (D-Frederick) said that the vote was "a pretty strong signal" of the general sentiment of the legislature on tobacco taxes (Associated Press 1999).

TOBACCO TAX ADVOCACY FROM 2001–2004

Progress began to look more promising after Dr. Leslie Beitsch began serving as Health Commissioner in the summer of 2001. Doug Matheny described the new galvanized atmosphere about tobacco issues, including the excise tax that began in the fall of 2001.

> The conventional wisdom in the public health community prior to that point in time was that we had to be realistic about our strength and that we could at best tackle one of these issues in a particular session. As the new health commissioner, Dr. Beitsch was adamant about reducing tobacco use. He just gathered together and mobilized the heads of state agencies as well as all the public health leaders and medical association and dental association. Everybody was up there, the whole works. From that point forward everybody just marched into the legislative session and a whole range of bills were introduced in all these subjects. (Matheny and Miner 2004b)

In December 2001, Governor Keating's Task Force on Tobacco and Youth publicly called for a tobacco tax increase in Oklahoma in order to raise revenue for use prevention and cessation. Rep. Ray Vaughn (R-Edmond), the task force chairman, said that an increase of one dollar per pack would elicit a 30 percent reduction in youth smoking. Ken Nance, then representing Brown & Williamson and Lorillard, told the Associated Press that the tax increase would be ineffective, stating that "the more you increase taxes on tobacco, the more the Indian smoke shops love it. It just increases business of Indian smoke shops" (Associated Press 2001). Dan Sorrels, former administrator for the Oklahoma Medicaid Program, also recalls the difficulties with increasing excise taxes in 2002.

> The state Medicaid program was facing a something on the order of 80 million dollar hold last year in the budget . . . going into 2002. The tobacco tax if passed and matched with federal Medicaid dollars would have more than filled that hole and because the Petroleum Marketers Association fought our efforts in the 2002 session so vigorously we were not able to get that legislation passed and because of that the state had to cut Medicaid services to thousands of Oklahomans . . . they are having a direct effect on that legislation.
>
> It is an incredibly strong lobby. It is the reason that we don't have a tobacco tax at this point in time is because of that association . . . because it is a retail association that is tied to the Chamber of Commerce and the State Chamber of Commerce and because it is also petroleum association and tied to petroleum interests in Oklahoma they are extremely powerful lobbying.
>
> I have talked to their lobbyists and been in meetings with their lobbyists with their legislators . . . I know that it is there. (Sorrels 2004)

By mid-January 2002, Mike Thornbrugh and Jim Hopper were publicly announcing opposition. Hopper, as president of the Oklahoma Grocers Association, unsurprisingly stated, "We're opposed to such a tax" (Associated Press 2002a). Thornbrugh, as government relations manager for QuickTrip (and House member from 1994 to 2000), declared that "The citizens of Oklahoma aren't thumping their chests demanding this large increase in their taxes," and that Oklahoma smokers could be driven to tribal smoke shops, bordering states, and the Internet. He added that some smokers might steal cigarettes or "orchestrate black-market tobacco schemes" (Associated Press 2002a). Richard Barnes, the chairman of the Oklahoma Alliance on Health or Tobacco, replied that a one-dollar increase, from the current 23 cents to $1.23, could raise $185 million for health care and tobacco prevention programs. Rep. Vaughn, the proposal's author, followed with the statement that 63 percent of Oklahomans polled were supportive of a one dollar increase as long as the money was designated for health issues (Talley 2003). Philip Morris contributed its resources for a public relations campaign like they had implemented to defeat a tax proposal in 1999. Oklahoma voters were called en masse and, if determined that the recipient was indeed an Oklahoma voter and appeared to oppose the tax increase, encouraged them to call their legislator, the name and phone number of whom were graciously provided by the caller (Philip Morris 2002).

In February 2002, Dr. Beitsch publicly declared, "Oklahoma's best and possibly only way to pass a one dollar per pack increase in the cigarette tax would be to put the measure to a statewide vote in November" (Borelli 2002). Beitsch further noted that "state lawmakers are out of step with the wishes of the public when it comes to tobacco control and prevention." Beitsch doubted the possibility of legislators raising tobacco taxes in the 2002 legislative session (Borelli 2002).

The 2002 legislative session passed without any headway on a tobacco tax increase. The issue was again proposed by Vaughn in 2003 via House Joint Resolution 1010, which called for a one dollar increase in the excise tax, but specified that the total tax per pack would not exceed $1.23 (Philip Morris 2002). Public support was announced by T. J. Brickner, M.D., Chairperson of the Oklahoma Health Care Authority and Tom Coble, president of the Oklahoma Association of Health Care Providers. Sen. Ben Robinson (D-Muskogee) stated that he had been persuaded to support the new tax increase due to the effect it might have on decisions to start smoking. However, the senator warned that "I will not support this unless there is a compact with the Indian tribes," reflecting the criticisms of tobacco advocates that traditional retailers would be unfairly treated if the tax resulted in disparity between their products and the tribal smoke shops (Wedel 2003).

An important challenge to the 2002 and 2003 tax increase proposals was State Question 640, which was passed by a statewide vote in 1992 and amended the Oklahoma Constitution such that revenue bills must attain a three-fourths majority vote in both houses or else go to a statewide referendum. Sen. Cal Hobson (D-Lexington), President Pro Tempore, admitted that this requirement would make the tobacco tax increase difficult to get through the legislature. Nonetheless, he remained an ardent supporter. "The time has come for this state to become mature about this killing instrument," he stated. "I don't want to be overly dramatic, but the fact is other states have gotten their act together on this issue. We have not and I'm determined we're going to before I get tossed out of here" (Carter 2003b). The President Pro Tempore also reiterated Sen. Robinson's concerns about tribal compacts, noting that 43 percent of all tobacco sales in the state occurred at tribal outlets, and asserting that:

> Without the support of the governor and a compact with the tribes, we have no chance for a significant cigarette tax increase passed by three-fourths of the Legislature . . . but I think with his help and cooperation of the tribes—who

> understand that they would benefit from this additional revenue—we have a real chance . . . the import point is that we allow the tribes, in cooperation with them, to maintain the current advantage that they have. (Carter 2003b)

Hobson also noted that in 1987, Oklahoma's 23 cent cigarette tax was the tenth highest in the nation, but was now ironically the tenth lowest in the nation. Meanwhile, the *Tulsa World* reported a survey of 500 registered voters in Oklahoma by Wilson Research Strategies that showed a 73 percent support rate from Republicans and 69 percent support rate from Democrats (English 2003). While Governor Henry began meeting with tribal leaders about negotiating a new compact, Thornbrugh told the *Tulsa World* that he did not believe that such a renegotiated compact would be successfully made, and reiterated the position that "our philosophy as a company is that if a tax is equal and fair across the board, it be remitted by all. That is fine. We will compete. When you have an unfair competitive advantage like the tribes do, we will oppose it" (Hoberock 2003).

While the tobacco tax failed to get to a House vote during the 2003 session, the issue was revived at the start of the new session in January 2004, consisting of a one dollar increase proposed by the Oklahoma Alliance on Health or Tobacco, under the leadership of Joy Leuthard, Director of Health Care Policy and Research for the Oklahoma State Medical Association. Dr. Mike Crutcher, the new Commissioner of Health after Dr. Beitsch returned to Florida, stated that previous failures of the tax increases were at least partly due to tobacco industry lobbying, and stressed that health advocates "need to be very persistent in our efforts" (Talley 2004b). Doug Matheny emphasized that progress on the tax issue was not a failure in 2002 and 2003, but simply that the momentum from Dr. Beitsch's 2001 initiatives took time to culminate in conjunction with new tribal compacts.

> Really in our minds there was not a tax effort last year [2003 legislative session] that failed and a tax effort this year (2004 legislative session) that succeeded, there was an initiative that was launched October 1 2001 that took three years to accomplish. All of these leaders that were going to be involved in this campaign always recognized there was no way that we were going to get an increase in tobacco tax until the compacts were renegotiated. The good news was the compacts were up for renewal in 2002 and so we were right around the corner from a 10-year expiration date. It was an opportunity to revisit them. Sure enough the governor did not renew the old compacts. And then we got into the 2002 elections and Governor Keating didn't get them renegotiated before leaving office. The new governor finished it up. (Matheny and Miner 2004b)

Meanwhile, the tobacco lobby was holding its ground as much as possible. Sen. Cal Hobson (D-Lexington) noted in February, during the new 2004 session, "the tobacco lobby has not gone away . . . there is a reason why the only states with lower tobacco taxes than Oklahoma are generally in tobacco-producing states" (Jenkins 2004a). Jamie Drogin, a spokeswoman for Philip Morris, replied that the company was concerned about illegal smuggling in states where tobacco taxes are raised, stating that "there has been an increase in illegal or contraband activities in relation to counterfeit cigarettes . . . it's very easy for kids to get cigarettes" (Talley 2004b).

Governor Brad Henry emphasized the positive impact of the tax revenue being designated for health care related expenses. His position was recorded in the *Journal Record Legislative Report* with the statement that:

> As I have said many times before, I am opposed to across-the-board tax hikes, but I would consider a targeted tax measure that earmark funding for a specific purpose. I would be supportive of a cigarette levy if the proceeds were specifically earmarked for health care initiatives such as a cancer center, trauma care, and medical coverage for the uninsured and youth tobacco prevention programs. I view the cigarette levy as more of a user fee that is only paid by those who use tobacco. (Carter 2004b)

The new tax proposal that Governor Henry endorsed was introduced as House Bill 2660 by Rep. Larry Adair (D-Stilwell) and Sen. Cal Hobson (D-Lexington) and called for a 77 cent per pack increase in the excise tax with an accompanying repeal of the 4.5 percent sales tax (25 cents per pack), culminating in a net 52 cent increase. Under the new tribal compact agreements that the governor had negotiated, while the tribal smoke shops had to collect the entire tax, half (26 cents per pack) would be returned to the tribes for health care initiatives. House Minority Leader Todd Hiett (R-Kellyville) complained that under the new tax plan, the tribes would enjoy an actual advantage even greater than they previously had, stating that "they're saying this closes the gap, but really what you're doing is adding a 26-cent disparity . . . you had a 17-cent disparity" (Carter 2004b). Rep. Heitt claimed that while the return is supposed to be used for tribal health care expense, and not funneled back to the smoke shops, the tribes will actually use the money received to free up other funds to aid the retailers. He was adamant that the new proposal enhanced an unfair tribal advantage.

> Money's all fungible . . . it's permission to use their sovereignty, basically, to sell cigarettes . . . when you look at the total package, this 26 cents is still going back into the tribes and so that's 26 cents that the tribes can give their operators in "cushion," advantage over non-tribal smoke shops—an additional 26 cents on top of the 17 cents. And they may not use the full 43 cents. They may not have to. They may use half of it—use 21 cents and pocket 22 cents into the

> tribe. They'll do whatever they'll have to do to bring all the market. It's not like they're not smart operators. (Carter 2004b)

Governor Henry disagreed, and claimed that the new tax proposal, along with the renegotiated compacts, would actually reduce the tribal competitive disparity by 60 percent. He sent a letter to Hiett and other legislators, including his Lieutenant Governor, Mary Fallin, detailing the dynamics of the legislation and compacts and attempting to address the concerns about tribal advantages and potential lost sales-tax revenue to cities and counties, which cannot impose a sales tax on items not sales-taxed by the state (Carter 2004b).

By the time the new House bill passed the Revenue and Taxation Committee by a 5–4 vote in February, tobacco lobbyist allies were on the scene to make negative public pronouncements. Jim Hopper, speaking for the Oklahoma Grocer's Association, reiterated the claim that the bill would give tribal smoke shops a greater competitive advantage. Mike Thornbrugh, representing QuickTrip convenience stores of Tulsa, said that the advantage to Indian operations would increase from $4.10 per carton to $5.55 per carton. Apparently dismissing the Governor's emphasis on having renegotiated more advantageous compacts, he stated "You can't fix this though legislation. It has to be fixed through compacts" (Associated Press 2004).

By March, House Speaker Larry Adair (D-Stilwell), who was attempting to garner support for the tax bill on the House floor, found that the vote count was stuck in the 30s (a bill must receive 51 votes to pass the House). Republicans were warning that the bill would not be successful as currently structured, but that there might be a compromise solution that would gain minority support. Lt. Governor Mary Fallin and House Minority leader Todd Hiett (R-Kellyville) announced that their caucus would be willing to support the tobacco tax bill if it included a provision eliminating the state capital gains tax. Fallin revealed a proposal to phase out the state's 6.65 percent capital gains tax over three years, which was reputed to be the 11th highest in the nation, which the Lt. Governor described as "a very oppressive tax that's on our books right now that's caused corporations to leave our state, has caused job loss within our state, has caused CEOs to leave our state" (Carter 2004e). Hiett announced that "if the governor is interested in a tax reform package, we obviously are very interested in looking at that . . . we have said for years that we need to reform our tax structure, restructure our taxes in the state of Oklahoma" (Carter 2004e). The Minority leader noted that many Republicans had made explicit pledges to oppose all tax increases, but if the increase in cigarette taxes was offset by reductions in the capital gains tax, then many Republicans could support such a combined measure without being susceptible to accusations of having reneged on their words. "If a package is revenue-neutral, that obviously is not an overall tax increase," he explained, "that is a tax restructuring and would not be a violation of the pledge" (Carter 2004e).

While the new bill gained some minority (Republican) support and passed the House 76–23, some Republicans were still unhappy about what they saw as uncertainty in the structure of the Indian compacts and the excise tax increase, and the floor substitute that left out substantive language until the compacts were finalized. Rep. John Wright (R-Broken Arrow) said, "Essentially, we're voting on a blank sheet of paper. I made the point that public policy should have disclosure, disclosure, disclosure, and there certainly isn't any disclosure going on tonight, because the details of the tax increase are hidden, hidden, hidden" (Carter 2004g). Rep. Richard Philips (R-Warr-Acres) was similarly agitated.

> This is not a shell bill. This is a $180 million tax that the tax has been taken out because they couldn't get enough of you to vote for it, so they met you in caucus and they met us in caucus and they said, OK, we're going to take the tax out, that way you can vote for the bill. The tax is still in there. You just can't see it. It is. Trust me, it's there. And it's coming back. Kill the bill now. Kill it now. . . . It's really a $180 million tax increase and then every quarter we write a check that's annualized in the amount of roughly $55 [million] to $60 million to the Indian tribes. And the Indian tribes can use that money in any way, shape or form that they want to. (Carter 2004g)

Further agitation came from a proposal by Rep. Bill Mitchell (D-Lindsey) to use $70 million of the anticipated $130 million raised by the new tax to provide a 7 percent pay raise to all state employees. Rep. Bill Graves (R-Oklahoma City) complained that "we're going to give an across-the-board pay raise and apparently we're going to do this through the latest jihad against cigarette smokers. And I don't think that's really fair. We're going to raise the taxes of cigarette smokers and really try to balance the budget on their backs" (Carter 2004g). Rep. Vaughn was concerned about keeping the revenues devoted to health funding, stating that "this is the same thing we did with the tobacco settlement. We should put this money aside and use it for the purpose for which it was intended, which was to remediate the health, the horrible health, of our citizens in the state of Oklahoma" (Carter 2004g).

Also in March, Wilson Research Strategies, a marketing research and management-consulting firm that had been conducting quarterly surveys of Oklahomans on key issues since 2002, released results from their poll of 300 likely voters in February regarding the proposed cigarette tax increase. With a margin of error of 5.6 percent, 64 percent of the respondents said that they would support an increase if it were linked to health care initiatives as proposed by the governor (Jenkins 2004b; Francis-Smith 2004).

Meanwhile, a poll composed and commissioned by the authors of this book and conducted by the University of Oklahoma Public Opinion Learning Laboratory (OU-POLL) in May 2004 found 75.5 percent of randomly surveyed Oklahomans as favorable to a 50-cent tax increase if it was linked to health care. Figure 5.1 indicates the proportions of Oklahomans' responses to proposed cigarette tax increases by amount and whether the funds would be designated for health care or for the general budget. The questions asked were: "To what extent would you favor or oppose an increase in taxes on cigarettes of [amount] per pack, with the revenue allocated in Oklahoma to the following: building a cancer research and treatment center, programs to reduce tobacco use among kids and help smokers quit, funding the emergency and trauma care center at OU Health Science Center, and bringing more Oklahomans access to Medicaid coverage with matching federal funds?" or "To what extent would you favor or oppose an increase in taxes on cigarettes of [amount] per pack, with the revenue allocated to the general budget?"

Further support for applying the tax initiative to health care came when Roxie Albrecht, director of Trauma and Critical Care Services at the University of Oklahoma Medical Center, addressed a joint House-Senate hearing in April. She warned that the trauma system at OU, the only level 1 trauma center in the state and designed to handle about 1,200 patients per year, typically treated about 2,800 and was suffering financial strain. Kansas had two centers in Wichita, Utah had two centers, and Nebraska three. House Speaker Larry Adair (D-Stilwell) noted that these states have populations smaller than Oklahoma's, and Albrecht stated that a state should have one trauma center per 1 million residents (Oklahoma has a 3.5 million population) (Carter 2003a).

In May 2004, Senate Republicans proposed a revised plan that would raise the cigarette tax by 99 cents per pack, and would also include a full repeal of the state capital gains tax, cuts in the estate tax and future reductions in the state income tax, as well as a proposal by QuickTrip that would allow non-tribal retailers to collect the same rates as local tribal retailers. Hobson said that the new proposal was not passable and continued to support the governor's 55-cent increase. Senate Republican leader James Williamson (R-Tulsa) expressed disappointment, calling Hobson's and the governor's stance a "partisan, take-it-or-leave-it plan," stating that the Republican alternative would provide "historic, job-creating tax relief" while deterring youth smoking. The American Cancer Society endorsed the new Republican proposal; Misti Rice, senior director of governmental relations, stated, "it would be a public health win that would reduce smoking and save lives" (Carter 2004a).

On May 20, 2004, after very nearly being defeated on a procedural vote, the House finally approved House Bill 2660 by a vote of 66–33, while the Senate followed suit the next day with a vote of 39–8, setting the stage for

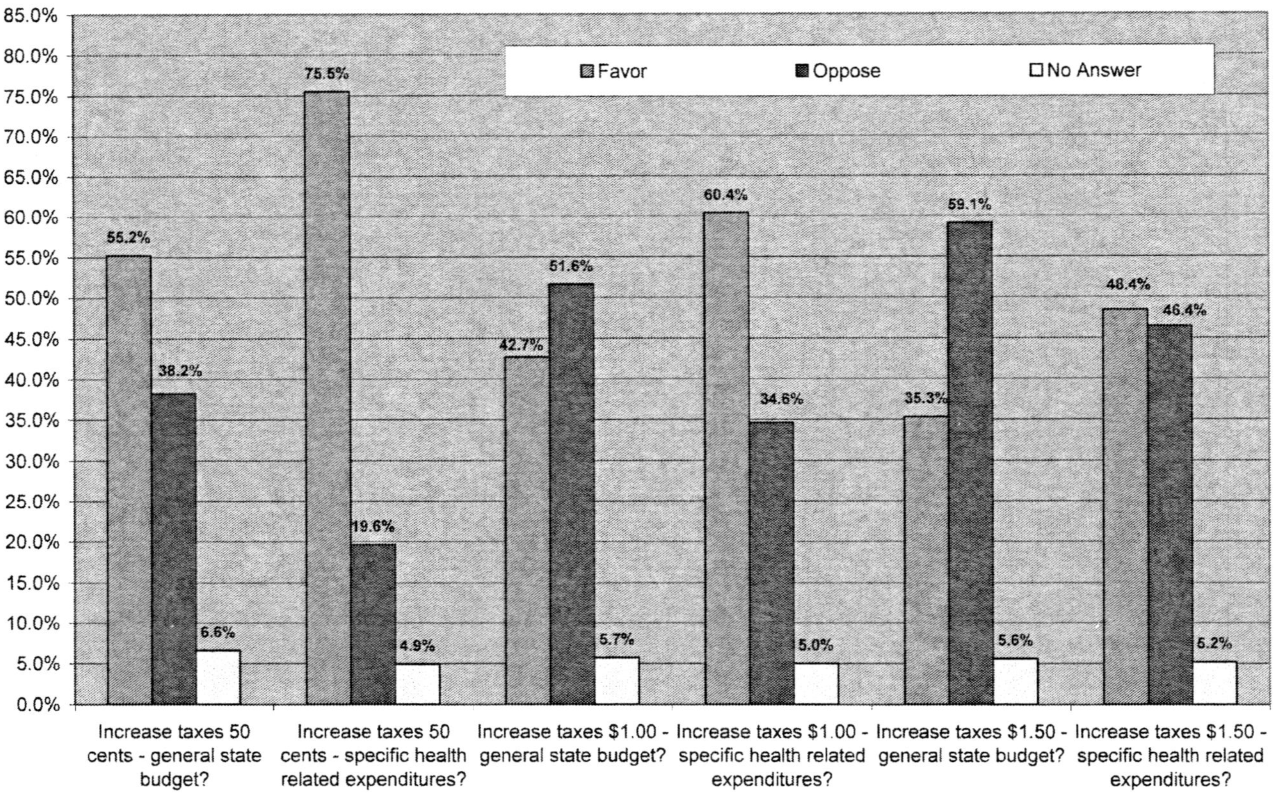

Figure 5.1. Bar chart of responses to May 2004 random sample Oklahoma statewide survey.

Source: Oklahoma Public Opinion Learning Laboratory, University of Oklahoma.

State Question 713, a referendum, to go to a vote of the people the following November. The bill would implement the governor's 55 cent tax increase, as well as eliminate the state capital gains tax, provide a tax cut for retirees, and permanently lower the maximum income tax rate from 7 to 6.65 percent (Carter 2004f; English 2004; Talley 2004a). Hobson declared "Don't let big tobacco kick you around any longer. They've owned this building for too many decades." Not everyone was as elated. Sen. Jim Dunlap (R-Bartlesville), said that the bill would be a step toward socialized medicine, and Sen. Randy Brogdon (R-Owasso) claimed that the $50 million designated for assistance to business to provide insurance to their employees was "corporate welfare" (Carter 2004h).

STATE QUESTION 713

While the Petroleum Marketers Association did not take a stand on State Question 713, the newly formed Tobacco Retailers Alliance announced in September 2004 that it would launch "a modest and truthful, unemotional campaign" to defeat State Question 713 on November 2, and expected to spend between $1 and $3 million. The group, led by president Joe Lane, an independent smoke shop owner, retained veteran public relations executive Dean Sims to manage their campaign. Sims was founder and chairman of Public Relations International, a Tulsa based firm (Carter 2004h).

In September 2004, former Tobacco Institute lobbyist Ken Nance, in addition to personally donating $250 to the Vote No on State Question 713 Committee, acted as signing treasurer for the committee, which received $2,683.90 from Philip Morris, USA that month (Oklahoma Ethics Commission 2004e). Originally organized on September 22, Nance was listed as "Temporary Chair" of the Committee, but appears to have remained so for the life of the committee (Oklahoma Ethics Commission 2004i).

During October, the committee received an additional $350,000 from Altria (Service Company of Philip Morris and Kraft Foods), $50,000 from the Cigar Association of America, $25,000 from Conwood Company LP (manufacturer of smokeless tobacco products), $1,000,000 from Philip Morris USA, $200,000 from the retailer QuickTrip, $200,000 from R. J. Reynolds, and $100,000 from United States Tobacco (UST) Public Affairs. Additionally, the committee received $128,201.09 of "In Kind" contributions from UST, Conwood Company LP, Lorillard, Philip Morris, and R. J. Reynolds for such services as "voter registration phone calls," "staff services," "printing," "postage," "consultant & employee time & expenses for work done on behalf of committee," "study on revenue implications of Q713 on OK Municipali-

Table 5.1. Oklahoma Ethics Commission, Form C-1R (November 10, 2004).

Media production and radio and TV buy	$396,000.00
Rental and installation of telephone system	$200.00
Web site production	$640.00
Billboard graphics	$650.00
TV and radio production, testing broadcast	$428,000.00
Campaign management	$10,000.00
Web site	$150.00
Management and organization of campaign	$100,000.00
Campaign management	$10,000.00
Development and implementation of media plan	$10,000.00
Development of retail collateral material	$11,000.00
Coordination of mail and phone program	$5,000.00
Development and implementation of statewide email	$15,000.00
Electrical computer, telephone work in campaign	$354.91
Television buy, radio buy, final production for tv and radio	$459,000.00
TV buy	$120,000.00
Web site	$350.00
Office expenses, furniture rental, traveling expenses, phone software	$1,499.26
Daily tracking poll	$60,000.00
Production and postage on mail piece	$77,865.00
Get out the vote recorded calls	$78,000.00
Office expenses/sign production	$679.07
TOTAL EXPENDITURE, October 1, 2004 to October 31, 2004	$1,784,388.24

Source: Campaign Contributions and Expenditures Report, Vote No on State Question 713 Committee, Schedule E.

ties," "employee time dropping off petitions in retail outlets," "public relations consultation employee salary & expenses" and "public relations consultation" (Oklahoma Ethics Commission 2004c). The committee reported an array of expenses for campaigning activities during October 2004, including the following (Oklahoma Ethics Commission 2004d).

Funding and expenditures for the pro-713 committee were modest compared with the competition. Organized on July 15, 2004, the committee Citizens for a Healthy Oklahoma was chaired by Oklahoma Alliance on Health or Tobacco Chair Joy Leuthard (Oklahoma Ethics Commission 2004h). During August, the committee received $50,420 from private contributors and $161,500 from organizations including the William K. Warren Foundation, the Broken Arrow Medical Center Foundation, the Oklahoma State Medical Association, Oklahoma Oncology Inc., and OU Medical Center. During the same month, the committee made media purchases totaling $125,035 (Oklahoma Ethics Commission 2004b). In September, an additional $423,000 was collected from contributors including Stillwater Medical Center, the Oklahoma Hospital Association, the Oklahoma Association of Health Care

Providers, and several other hospitals and medical providers. During the month, $388,155 was spent on legal services, advertising expenses, and campaign management services (Oklahoma Ethics Commission 2004b). During October, various medical providers, as well as Walgreen Company and the Greater Oklahoma City Chamber of Commerce contributed an additional $261,000. Reported expenditures were comparable, and included items listed as advertising, professional services, and wages (Oklahoma Ethics Commission 2004a).

After a barrage of political advertising on both sides (figures 5.2 and 5.3), on November 2, 2004, State Question 713 was passed with 53 percent of the popular vote. The referendum eliminated the 25-cent state sales tax on cigarettes and increased the excise tax by 80 cents, from 23 cents to $1.03, for a net tax increase of 55 cents (Carter 2004i). The basis of the successful win was constant outsider pressure by the Oklahoma Health Commissioner and organized health groups. As long as they engaged in insider lobbying tactics including in the 2002 legislative session, the chances of any tobacco tax increase were quite remote.

Despite the success of State Question 713 and the raising of Oklahoma's excise tax on cigarettes, the state still prohibits and preempts local munici-

Figure 5.2. Advertisement from the Vote No on Question 713 campaign.

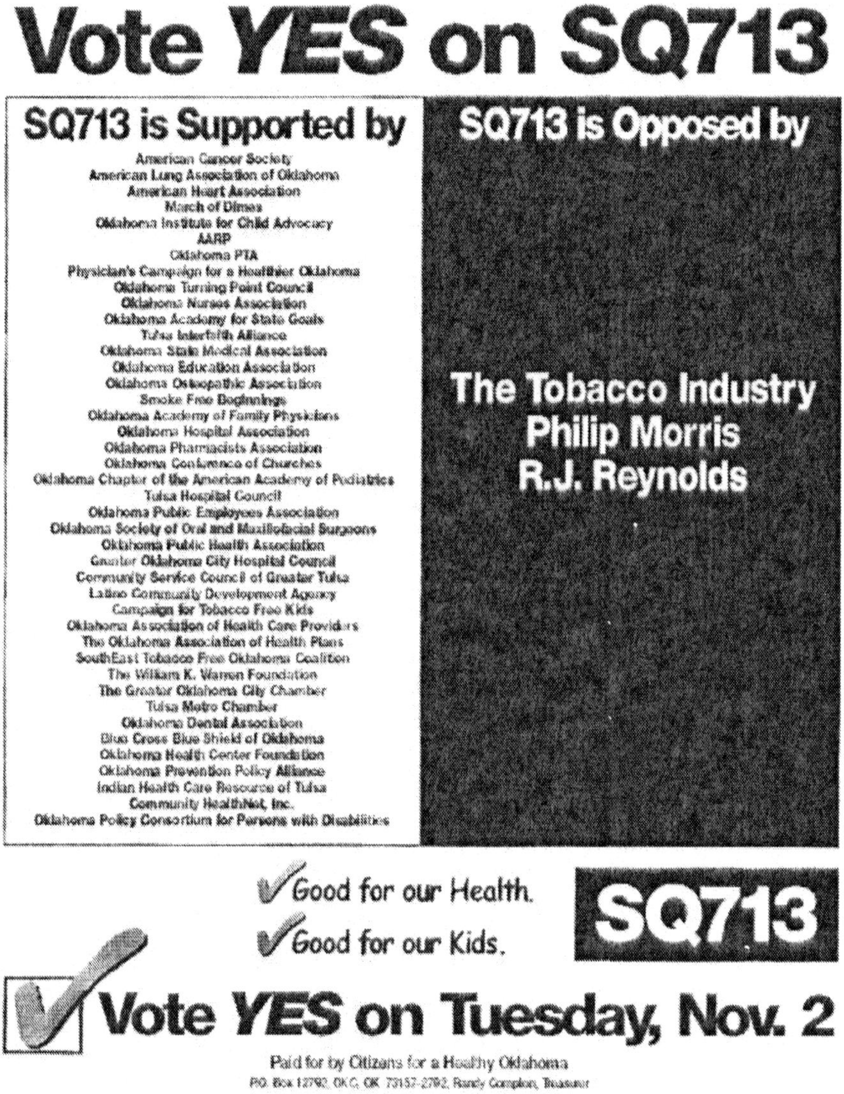

Figure 5.3. Advertisement from the Vote Yes on State Question 713 campaign.

palities from taxing tobacco products, as stated in Oklahoma Statute Title 37, section 600.10, implemented with the 1994 Youth Access to Tobacco Act. Local municipalities are allowed to impose local sales taxes on all items subject to state sales tax, but as of January 1, 2005 the list of goods and services subject to state sales tax did not include tobacco, since the state sales tax was repealed in favor of a higher excise tax (Carter 2004d).

Chapter Six

Aftermath

Clean Indoor Air Post–2003

Since the enactment of SJR 21 and State Question 713, few victories in Oklahoma's legislature have had enough muscle to pave the way for more effective anti-tobacco legislation. But there has been some effort. In 2007, SB 473 prohibited smoking at zoos (Oklahoma State Senate 2007). As of January 2009, all Oklahoma-sold cigarettes must self-extinguish when not being smoked in order to prevent accidental fires. Cessation coverage for state employees increased along with an expansion of Medicaid cessation services (Oklahoma Insurance Department 2007; Oklahoma Health Care Authority 2010). Additionally, 2003's Certified Healthy Business Program, supported by a number of partners, includes extra certification points for being 100 percent tobacco-free and has already certified 340 businesses (Oklahoma State Department of Health 2009). By March 1, 2006, restaurants were required to have completed a fully enclosed, separately ventilated, and negative air draft smoking room if they decided to continue to allow smoking inside (Oklahoma State Senate 2003a). Passed in November of 2004, HB 2660 increased the excise tax on a 20-pack of cigarettes by 80 cents to $1.03 per pack (the average is $1.42 nationally as of July 1, 2010; Campaign for Tobacco Free Kids 2010).

In 2008, the Tobacco Settlement Endowment Trust launched the "Tobacco Stops with Me" campaign, including billboards, television ads, newspaper inserts, and a Facebook page (Oklahoma Tobacco Settlement Trust 2008). On July 1, 2008, the Attorney General's office banned tobacco, both inside and outside (Oklahoma Attorney General 2008a). Many colleges and universities around the state were either smoke-free or were expected to become so after 2010 (American Lung Association 2010a). However numerous, these new initiatives have not appeared to impact the quality of air in the state; smoking is still allowed in bars, restaurant smoking rooms, and casinos across the state.

As described in chapter 4, the years-long struggle toward a smoke-free state culminated in the first substantial progress in 2003 with SJR 21, which prohibited smoking in interior workplaces and public places, with the exception of bars and taverns (and Native American-owned businesses, which we will discuss later). On May 7, 2010, Governor Brad Henry signed House Bill 2774, which authorizes a 50 percent reimbursement for a restaurant's cost of building smoking rooms, if the proprietor dismantles the room (voluntarily) by January of 2013. Authored by Representative Kris Steele (Speaker Pro Tempore/Republican), the bill is not a mandate. Instead, it lays the groundwork for a possible smoking ban in the future by removing the argument stating that restaurants cannot afford to dismantle their newly built smoking rooms (Klein et al. 2010; Rogers and Peterson 2008).

Efforts toward a completely smoke-free state continue to be hampered by, among other organizations, the ORA. The ORA's lobbyist, Benny Vanatta (who also represents Swisher International, Inc., makers of cigars and smokeless tobacco), singled out the unregulated, Native American-owned casinos in the state as the major factor in ORA's opposition to a statewide smoking ban. Were such a law passed, restaurants inside casinos could compete easily for smoking patrons, Vanatta argued in *The Oklahoman*, stating "That's an issue that we don't know if it is solvable" (Estus 2009). Also, some restaurant owners feel they are doing a service to their patrons by keeping a smoking room available. The owner of Cattlemen's Steakhouse in Oklahoma City, Dick Stubbs, told *The Oklahoman*, "We get a lot of comments from smokers saying thank you for having a place for them" (Colberg 2010). Other restaurateurs are more supportive and vocally in favor of anti-smoking measures. Ken Selby, owner of Mazzio's and Oliveto, converted his restaurants into smoke-free establishments over a decade ago and told the *Tulsa World*, "Our employees' exposure to harmful secondhand smoke was eliminated and we experienced no decline in our revenues" (Hoberock 2010).

During the legislative session of 2010–2011, House Bill 2135 exhibited the strongest promise to turn the tide of tobacco control in over two decades (*Tulsa World* 2011; *The Oklahoman* 2010). Authored by Speaker of the House Kris Steele, the bill sought to reverse the preemption language of the 1987 *Smoking In Public Places Act* by giving local governments authority to enact stronger clean indoor air laws than those mandated by the state. In response to increasing public support for HB 2135, tobacco companies increased their Oklahoma presence, hiring a record 13 representatives to lobby for the industry (Oklahoma Ethics Commission 2011; Estus 2011).

The bill escaped committee with a 9–5 vote, but died on the House floor in March of 2011. Though a groundswell of support had carried the bill along to reverse decades of preemption, the tobacco industry once again demonstrated its ability to influence legislation in its favor (Stearns 2011). In 2012 while passing in the Oklahoma House by a two to one margin, a companion bill to repeal state preemption of stronger local clean indoor ordinances failed to make it out of the Senate Committee on Health and Human Services (Hertneky 2012). Committee Chair Senator Brian Crain (R-Tulsa) killed the bill by refusing to allow it to be heard (Hertneky 2012). Defending his action, Senator Crain stated, "he [Senator Crain] believes this bill was an incremental approach to banning smoking, and he's concerned we're slowly taking away the rights of smokers 'without them being aware of what's going on'" (Hertneky 2012). The death of HB 2135 in 2011 and HB 2267 in 2012 was a significant victory for corporate tobacco forces in Oklahoma.

ATTITUDES TOWARD CLEAN INDOOR AIR LEGISLATION

Popular opinion helped spur the passage of SJR 21, the 2003 anti-tobacco legislation that prohibited most indoor smoking (Spivak and Givel 2005). Since then, public opinion further evolved on the issue of smoking in public places. Among respondents to the 2008 Oklahoma Adult Tobacco survey in 2002, 46 percent of those surveyed said smoking should not be allowed in any public places. The survey did not ask this question in the same format in subsequent years, but instead asked for opinions about smoking in specific public places.

By 2008, respondents showed themselves to be much more critical of smoking in public, although their opinions appeared to vary substantially between the types of establishments. Bars, for example, did not elicit the strong anti-smoking sentiments as toward tobacco use in restaurants and shopping malls.

Oklahomans responding to the question in the 2008 Oklahoma Adult Tobacco Survey, "About how often do you go to a bar?" were categorized by: more than once a month (8.0 percent), less than once a month (15.1 percent), and never (76.4 percent). A small percentage did not answer the question or did not have an answer to the question. Those respondents who never visited bars disapproved of smoking in bars at a rate of 45.6 percent, while those who went to bars less than once a month disapproved at a rate of 29.0 percent. Frequent bar-goers (more than once a month) disapproved of smoking at a rate of 31.6 percent.

A little over 10 percent of respondents who never visited bars believed that smoking should be allowed in all areas of a bar. For those going to bars more than once a month, 41.5 percent believed that smoking should be allowed in some areas and 26.1 believed smoking should be allowed in all areas of a bar or cocktail lounge. Figure 6.1 shows the distribution of public opinions about allowing smoking inside bars and cocktail lounges, categorized by frequency with which respondents visited such establishments.

When it came to places patronized regularly by respondents, such as restaurants and shopping centers, stronger anti-smoking attitudes emerged. Between 2002 and 2008, the percentage of respondents who believed smoking should not be allowed at all in restaurants rose from 56.9 percent to 67.0 percent. The national average, using 19 states' Adult Tobacco Survey data, shows that 65.5 percent of adults support complete smoking bans inside restaurants (McClave et al. 2010).

Additionally, support for smoking in some areas of restaurants (i.e. smoking rooms) dropped from 38.5 percent in 2002 to 30.3 percent in 2008 (see figure 6.2), further buttressing the call for clean indoor air. Twenty-six states now prohibit indoor smoking in restaurants—there are exceptions to this rule based on state laws, with some states being 100 percent smoke-free and some

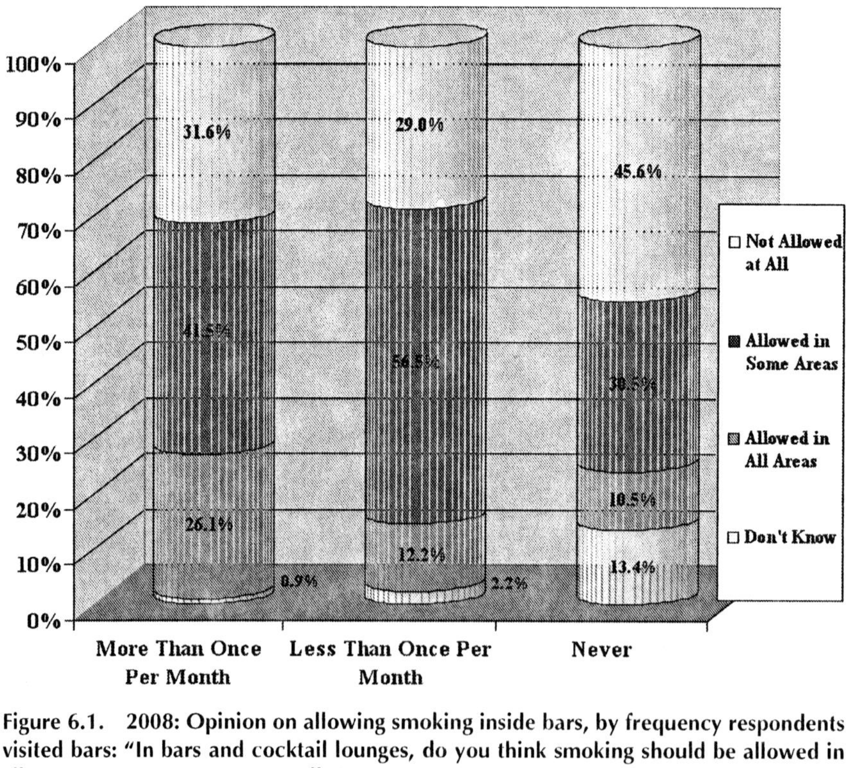

Figure 6.1. 2008: Opinion on allowing smoking inside bars, by frequency respondents visited bars: "In bars and cocktail lounges, do you think smoking should be allowed in all areas, some areas or not at all?"
Source: 2008 Oklahoma Adult Tobacco Survey, Oklahoma Department of Health.

not (American Lung Association 2010b)—while Oklahoma continues to allow smoking inside specially built smoking rooms of dining establishments. Unfortunately, House Bill 2774, with its provision to reimburse one half the cost of dismantling smoking rooms has failed to elicit a positive response in the legislature, probably due to both its lack of mandatory participation and the fact that the special rooms had been built only a few years ago.

From 2004 to 2008, Oklahoma Adult Tobacco Survey respondents who thought smoking should not be allowed at all in shopping centers rose from 69.4 percent to 74.4 percent (see figure 6.2). The data show that, at least for establishments like malls and restaurants, public support has grown increasingly in favor of clean air. In comparison, support for clean air in bars and taverns is not as strong, with 42.3 percent of respondents in favor of no smoking at all inside bars (in figure 6.2, opinion on smoking inside bars is collapsed into one category of respondents).

By 2008, 69 percent of respondents believed that breathing others' cigarette smoke was "very harmful" to their own health. The seriousness of this

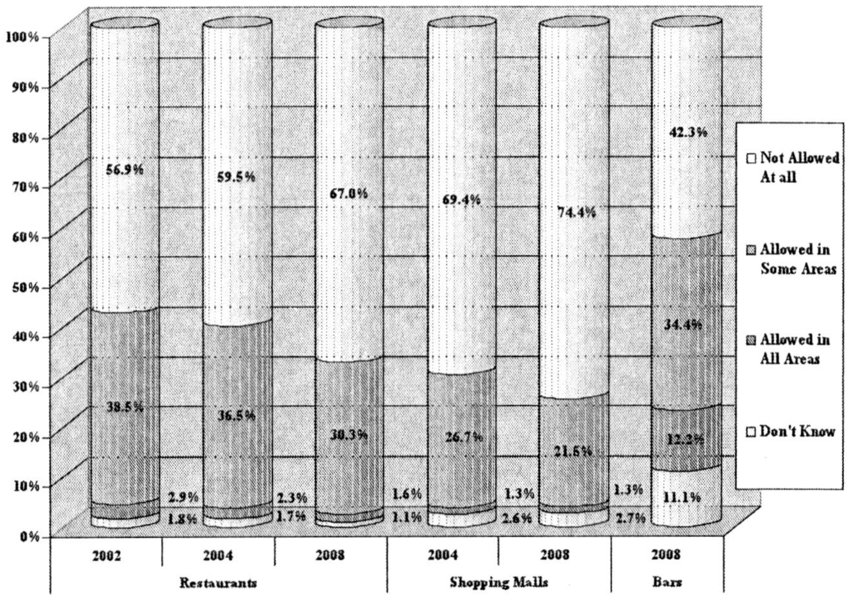

Figure 6.2. Opinions about smoking inside restaurants, shopping malls, and bars: "In indoor shopping malls, indoor dining areas, and in bars and cocktail lounges, should smoking be allowed in all areas, some areas, or not at all?"
Source: 2002, 2004, 2008, National Adult Tobacco Survey (NATS): Centers for Disease Control and Prevention (CDC). See http://www.cdc.gov/tobacco/data_statistics/surveys/nats/index.htm.

threat translated into greater concern for the quality of air breathed by respondents in the public arena. In addition to asking about attitudes toward clean air in restaurants, the Oklahoma Adult Tobacco Surveys also examined attitudes toward clean workplace air. SJR 21 eliminated tobacco use in most public places and indoor workspaces in 2003 (Oklahoma State Senate, 2003b), but majority support existed even prior to the legislation's passage. As early as 2002, only 30 percent of respondents believed smoking should be allowed in some indoor work areas. This percentage dropped to 18 percent in 2008, while support for no smoking at all in indoor workplaces rose from 64 to 75 percent over the same six-year period (see figure 6.3).

Respondents also reported their own employers' workplace policies. In 2002, before SJR 21, 68 percent of respondents' employers banned indoor smoking completely. By 2008, this percentage had risen to 80 percent. The proportion of employers allowing smoking in all work areas, or having no policies at all, stayed well below 7 percent during all three waves (2002, 2004, 2008) of the Adult Tobacco Survey (see figure 6.4).

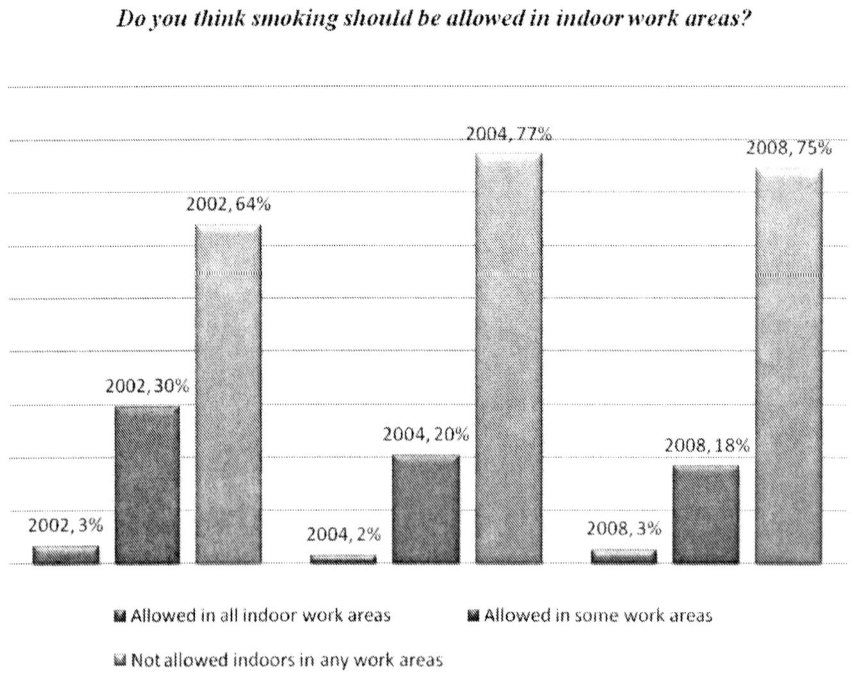

Figure 6.3. Opinions on smoking in indoor work areas from 2002 to 2008.
Sources: 2002, 2004, 2008 Oklahoma Adult Tobacco Surveys.

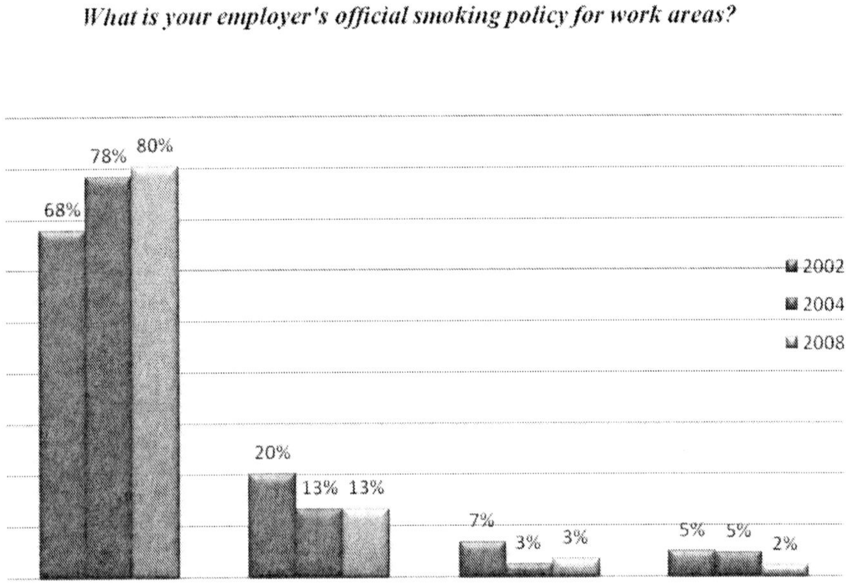

Figure 6.4. Employers' official smoking policy for work areas from 2002 to 2008.
Sources: 2002, 2004, 2008 Oklahoma Adult Tobacco Surveys.

THE TOBACCO INDUSTRY'S EFFECT ON THE OKLAHOMA LEGISLATURE, 2005 TO 2010

This section examines the influence of the tobacco industry and its lobbyists on the state government in Oklahoma between 2005 and 2010. Since its formation in 1991, the Oklahoma Ethics Commission has overseen lobbying activities and campaign contributions (Oklahoma Ethics Commission, 2010a, 2010b, 2010c). Among the Commission's mandated tasks is the registration of all lobbyists, a comprehensive list of principals by whom lobbyists are retained, all financial expenditures of lobbyists to members of the legislative, executive, or judicial branches of the state government, including state employees, and full disclosure of principals on whose behalf lobbyists undertake financial expenditures.

The Commission also requires campaign committees and political parties to disclose contributions. Committees must provide the amount of a contribution and to whom the contribution was given. To further explore the trends in tobacco industry contributions, the following terms are defined according to the Ethics Commission.

As defined by the Commission, a lobbyist is, first and foremost, someone hired by an interested third party to engage in lobbying (Oklahoma Ethics Commission 2010b). Lobbying involves the act of written or oral communication with a member of one of the three branches of government on behalf of some third party group (a principal) with the aim of providing input on policy or legislation. A lobbyist principal is defined as "any person who employs or retains another person for financial or other compensation to conduct lobbying activities on behalf of the lobbyist principal" (Oklahoma Ethics Commission 2010b).

A financial expenditure made by any committee, family, individual, or lobbyist to a political campaign committee is known as a contribution. This includes loans or forgiveness of loans, payment in cash, and purchased tickets to political fundraising or campaign rallies. The Commission also considers a contribution "anything of value made to and with the knowledge and for the benefit of a committee for use in a campaign, or for reducing the debt of a committee" (Oklahoma Ethics Commission 2010b).

A financial expenditure made by a lobbyist on behalf of a principal is called a gift. The Commission defines a gift further as "anything of value" exceeding ten dollars (Oklahoma Ethics Commission 2010b). This includes "soft money" (e.g., meals, entertainment, athletic event tickets) and "hard money" (e.g., a check or honorarium).

Lobbyists must file reports with the Commission twice a year, disclosing their financial activities during the previous six months. These reports must include the lobbyist's name and address, the name and position of the government official (and, if applicable, the official's immediate family member) with whom this financial activity occurred, the date of the activity, the nature and cost of the activity (or thing of value given), and the principal under whom the lobbyist is acting at the time the thing of value is given.

For the years 2005 through 2007, the calendar gift limit for lobbyists and lobbyists' principals was $300 per elected official. The Commission lowered this amount to a $100 limit per elected official, per annum, beginning in 2008 (Oklahoma Ethics Commission 2010a).

TOBACCO INDUSTRY LOBBYISTS IN OKLAHOMA 2005–2010

Registration of lobbyists and their principals comes from records housed at the Oklahoma Ethics Commission. Lobbying reports from 2005 are in paper form at the Commission's physical location inside the Oklahoma City Capitol building, while the years 2006 through 2010 are obtainable in an online database at the Commission's website (Oklahoma Ethics Commission 2010c). Because the Commission exists for purposes of transparency in government, any member of the public has access to financial data from campaigns and lobbying.

In the next sections, we report each lobbyist who registered under one or more tobacco principals during the years 2005–2010. Expenditures noted below are for gifts or in-kind contributions given to state policymakers from lobbyists, and not for campaign contributions.

Chad Alexander

Reynolds American hired Alexander as a lobbyist in 2011. Alexander currently also represents the University of Oklahoma and Oklahoma State University. He did not report any gifts given to legislators on behalf of Reynolds American in the first half of 2011.

Eric Donaldson

Donaldson registered as a lobbyist with U.S. Smokeless Tobacco, John Middleton Co. (a manufacturer of cigars and pipe tobacco), and Philip Morris from 2009–2010 but did not show any gifts made to legislators on behalf of these principals.

David Stephen Edwards

Edwards was hired by Xcaliber International in 2011 as a lobbyist, but did not report any gifts made to legislators on behalf of Xcaliber during the short time he has been retained by the company.

Elizabeth Giles

Registered as a lobbyist on behalf of Philip Morris from 2006–2008, Giles reported she gave no gifts to legislators under this tobacco principal during the period studied.

Richard Haselwood

Haselwood listed principal Reynolds American (R. J. Reynolds) in 2009 on his registration form, but did not report any gifts given on behalf of this tobacco company.

Richard Huddleston

Huddleston served as Chief Clerk of the Oklahoma House of Representatives from 1976 through 1983 (Humphreys 2000). He lobbied for U.S. Smokeless

Tobacco from 2001 to 2008 but was not financially active on behalf of this principal during the years studied. Huddleston listed his own agency, Huddleston Investments, Inc., as a principal on whose behalf he gave a thing of value to a policymaker between the years of 2005 and 2010. A loophole in the Ethics Commission guidelines allows lobbyists to make expenditures on behalf of their own agencies as principals, thus making these types of transactions untraceable to any other primary source except the lobbyist's company. Richard Huddleston passed away in March 2008 (Social Security Death Index 2010).

Courtney Lee Jones

Xcaliber International retained Jones as a lobbyist in 2011. Jones did not report any gifts given to legislators on behalf of Xcaliber during the first half of 2011.

James Milner

Like Haselwood, Milner registered as a lobbyist under Reynolds American in 2009 but did not report any financial activity. Milner also listed Conwood Tobacco as one of his principals in 2006 but reported no gifts.

Brian Nance

Brian Nance is the son of tobacco lobbyist powerhouse Ken Nance (Francis-Smith 2006). Brian Nance registered as a lobbyist with R.J. Reynolds/ Reynolds American, Inc. from 2006 through 2010. He did not report any expenditure in 2006, but was financially active on behalf of Reynolds from 2007 through 2010, according to his financial reports. In 2007, he reported expenditures for gifts to legislators in the amounts of $239, $820.79 in 2008, and $203.06 in 2009. During 2010, Nance reported spending $813.33 for gifts to legislators on behalf of Reynolds American, Inc.

Ken Nance

The tobacco industry acknowledged Ken Nance as an important force in the 1990s, and he continued to influence tobacco legislation in the 2000s, according to insider memos and documents (Spivak and Givel 2005). Nance served as an Oklahoma Representative from 1969–1977 (Spivak and Givel 2005). By the time of the current study, Nance had virtually disappeared from tobacco dealings. He listed Conwood Tobacco and Swedish Match (manufacturer of cigars and chewing tobacco) (Swedish Match 2010) as principals on his 2005 registration, but reported no financial activity on their behalf. Nance's termination from both Conwood Tobacco and Swedish Match was filed on January 13, 2006. He declared all his 2005 expenditures to legislators were made on behalf of the principal "Ken Nance," even though his name is not listed as a principal for 2005. Nance does not attribute further gifts to the principal "Ken Nance" in the years 2006 to 2010, nor does he register as a lobbyist for any tobacco companies after 2005. Other lobbyists who made gifts on behalf of their own lobbying companies listed their companies as principals on the registration forms; it is unclear why Ken Nance did not.

Richard Phillips

An Oklahoma Representative from 1993–1999 (Humphreys 2000), Phillips the lobbyist listed the Cigar Association of America as a principal from 2008–2010. Phillips did not report expenditures in 2007, but undertook financial transactions on behalf of the Cigar Association to purchase gifts for legislative members and the assistant director of the Alcoholic Beverage Laws Enforcement Commission (ABLE) during the years 2008, 2009, and 2010. He reported gift expenditures of $71.11 in 2008 and $253.27 in 2009. During the first half of 2010, Phillips reported $106.11 for gift expenditures.

Dianne Rasmussen

Rasmussen listed Conwood Tobacco as a principal in 2006, but did not report any gifts given on behalf of Conwood. Rasmussen owns her own lobbying company in Oklahoma City, Dianne K. Rasmussen LLC.

Tom Rogers

Tom Rogers served as an Oklahoma Representative from 1969–1977 (Humphreys 2000). Records at the Ethics Commission show that Rogers registered as a lobbyist for U.S. Smokeless Tobacco in 2009, Philip Morris from 2005–2010, and John Middleton Co. in 2009. However, he did not report expenditures on behalf of any of these principals during those years. Rogers's own lobbying agency, GRPA Agency Inc., is listed as a principal on whose behalf he contributed items of value to various legislative staff and members during the years studied. However, as the Ethics Commission allows lobbyists to report activity under their own agencies, these monies are not traceable past any principal other than Rogers's own GRPA Agency.

Robert Stem

Stem listed the Tobacco Retailers Alliance (an Oklahoma-based alliance) as one of his principals from 2005 to 2006 and Xcaliber International (an Oklahoma-based cigarette manufacturer) from 2007 through 2010 (Xcaliber 2010). He did not report any gifts given on behalf of either principal during the time period studied. In 2007 and 2008, Stem also lobbied for the Tribal Retailers of Oklahoma, a group of Native American-owned businesses with an understandable interest in cigarette excise taxation, since some members of the Tribal Retailers own tribal smoke shops. Stem did report gifts made on behalf of the Tribal Retailers, but these monies are not relevant to this report, since the Tribal Retailers of Oklahoma is not strictly tobacco-based businesses.

Benny Vanatta

A former Oklahoma legislator from 1981 through 1988 (Centre for Public Integrity 2006; Humphreys 2000) Vanatta represented Swisher International (Swisher 2010), a cigar company, during 2005–2010. Vanatta also represented the Oklahoma Restaurant Association during this same time period, long considered one of the tobacco industry's allies (Spivak and Givel 2005). He did not list any expenditures associated with Swisher International during the period studied, but was quite financially active on behalf of the restaurant association.

Nathan Thomas Webb

Retained as a lobbyist by Xcaliber International in 2011, Webb did not report any gifts made to legislators on behalf of Xcaliber International during the first half of 2011.

Don Williams

Williams, a former Oklahoma Senator (1989–1995) (Oklahoma Department of Libraries 2007a), registered as a lobbyist for Philip Morris from 2005–2010, U.S. Smokeless Tobacco in 2009, and John Middleton Co. from 2009 through 2010. He did not list any financial expenditure on behalf of these tobacco companies during the time period studied. However, similar to lobbyists Tom Rogers and Richard Huddleston, Williams lists his own company, WGRC Inc., as a principal for whom he lobbies. Reports at the Ethics Commission show a considerable number of gifts to legislators attributed to the principal WGRC Inc. during the time period studied. However, these monies are not traceable to a primary principal past this company. Williams also lobbied for the Oklahoma Wholesale Marketers Association (OWMA) from 2005 through 2010 (with the exception of 2009, when the OWMA did not retain any lobbyist) and has been vocal about the issue of excise taxes on cigarettes on behalf of this organization (Associated Press 2007). The OWMA's website features 2007 news about cigarette taxation and states that association members act as tax collectors by putting tax stamps on packages of cigarettes sold to retailers (Oklahoma Wholesale Marketers Association 2010). Clearly, this organization has a stake in cigarette taxation proposals.

CONTRIBUTIONS THROUGH TOBACCO INDUSTRY POLITICAL ACTION COMMITTEES

The total amount funneled into Oklahoma political processes during 2005–2010 from the tobacco industry equaled $75,875. Figure 6.5 provides a visual representation of the amounts contributed by tobacco PACs, per year. These contributions include the underwriting of formal balls, donations to campaign committees, and gifts or contributions to political parties.

Additionally, we examined the impact of tobacco contributions on Oklahoma legislators between 2005 and 2010 through a stepwise ordinary least square regression, assessing the extent to which political party, gender of a legislator, representation of higher percentages of Native Americans in a district, years in office, urban area representation, or political leadership was significantly correlated with political campaign contributions received from

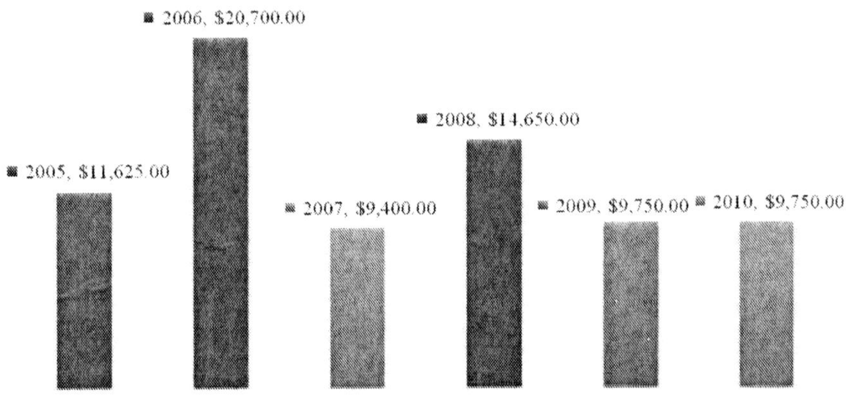

Figure 6.5. Contributions to political campaigns, political parties, sponsorship of balls, and other PACs by the tobacco industry PACs' direct contributions per annum.
Source: Oklahoma Ethics Commission Results.

the tobacco industry and legislators' pro (vs. anti) tobacco policy scores. This study was accomplished by a database compilation of 170 legislators, using the Oklahoma Department of Libraries' (ODL) yearly issues of *ABC: Oklahoma State Agencies, Boards, and Commissions* from the years 2005–2010. Every September, the ODL publishes a booklet containing comprehensive lists of all elected officials in Oklahoma government as of September 1, including all current members of the House and Senate. Some issues of ABC are online, while all of them can be obtained at the Jan Eric Cartwright Memorial Library in room B8 of the State Capitol building or in the Government Documents section at the University of Oklahoma Bizzell Memorial Library in Norman, Oklahoma.

Each House and Senate member was assigned to his or her corresponding party and gender, according to the Oklahoma State Legislature webpage (Oklahoma State Legislature 2010) and the Oklahoma Ethics Commission webpage (Oklahoma Ethics Commission 2010c). We also used these websites (along with the issues of ABC) to note which legislators had occupied positions of leadership at any point during their tenure from those who had never occupied a leadership position. Holding a leadership position includes filling Senate positions such as President Pro Tempore (highest ranking Senator), Whip (a party enforcer), Co-Assistant Floor Leader, or Caucus Chair. In the House, leadership positions encompass the titles of Speaker of the House (highest ranking Representative), Majority Whip, Minority Caucus Chairman or Minority Leader, among others. From the Oklahoma State Legislature webpage, we also assigned each legislator codes indicating whether he or she represented a district in the Oklahoma City or Tulsa areas (metro), versus all other areas (non-metro), as well as figures indicating the proportion of his or her district constituency that is Native American and Other Nonwhite (see Independent Variables, below). Metro areas include surrounding suburbs in accordance with the U.S. Census Bureau's definition of the Metropolitan Statistical Areas (MSAs).

Some former House members became Senate members, and vice versa, during the years studied. We remedied this issue by tallying the time spent in the House and the time spent in the Senate, coding the legislator as a Representative or Senator depending upon the seat they occupied for the most number of years. We also determined how many years each legislator held a seat, and if they exited their legislative position before the term limit. The Oklahoma Constitution specifies a legislator may not hold office for more than 12 years (Oklahoma State Constitution, Limitation of Time Served in the Legislature 1990). The 12 years need not be consecutive, but the time spent in service to both the Senate and the House (if a Senator becomes a Representative or vice versa) is added together and must fall under the 12-year limit.

Our analyses made use of independent third parties with a solid familiarity of tobacco issues in the Oklahoma legislature to rate legislators along a continuum, which represents a pro- or anti-tobacco industry stance. We began the process of selecting raters by making inquiries among representatives from health associations, grassroots tobacco control organizations, and journalists and legislative aides familiar with state legislators and tobacco control issues. Once we had a respondent express initial interest, we asked this first rater to suggest other willing individuals from these fields, adding the final two raters using this snowball method.

METHODS

The analyses contain two interval level dependent variables: (1) total campaign contribution amount from the tobacco industry and (2) tobacco policy score. In the multivariate models in which tobacco policy score is regressed, campaign amount is also an independent variable.

For the analyses, we examined direct campaign contributions only from tobacco PACs. By cross-referencing reports at the Oklahoma Ethics Commission filed by legislators' campaign committees and reports filed by tobacco PACs, we compiled financial contributions made to each of the 170 legislators from 2005 through 2010.

The second dependent variable is the tobacco policy score. Because our raters were familiar with legislators' voting habits and the tobacco control climate at the Capitol, we created the "Oklahoma Legislators Tobacco Attitudes Scoring Sheet" and implemented a Likert scale ranging from 0–10 to rate legislators' pro- or anti-tobacco stance. The scoring sheet included 172 legislator names (2 legislators were later discarded due to missing data from all 3 raters), party affiliations, districts represented, and years in office.

Instructions on the sheet asked the raters to circle a number from 0 to 10 to represent each legislator's attitude toward the tobacco industry. The Tobacco Attitudes Scoring Sheet explained the Likert scale of 0–10, describing a score of 0 as representing a legislator who is extremely anti-tobacco control, voting in a manner consistent with allowing the tobacco industry free rein, that is, "Laws should not attempt to curtail the tobacco industry's activities." The instructions explained that legislators consistently promoting policy that would regulate all aspects of the tobacco industry's activities should receive a score of 10, that is, "Laws should control the tobacco industry's activities." Raters were instructed to circle "N/A" for any legislator with whom they were unfamiliar or undecided, and also were instructed to indicate only one score for each legislator. We advised the raters that their answers would be kept confidential and anonymous, and results would be summarized, not individually reported. The raters worked independently of one another, without approving, supervising, or assisting with one another's surveys.

Reliability analysis produced a Cronbach's alpha of .82 for the three raters, with no increase in reliability when any of the three were removed. This reliability test confirms that the three raters, while providing some variation in their scores, tend to judge legislators similarly when assigning their ratings independently. In other words, no rater had significantly different opinions than the other two, although using three such judges provided variation with which to enhance the validity of each legislator's score. Ordinarily, a combined scale would be comprised of the sum of the three scores for a range between 0 and 30. However, we averaged them due to 30 cases having missing data on one or more of the raters. When we had a missing score, we averaged the remaining two, and if two were missing, we used the remaining individual score.

INDEPENDENT VARIABLES

The analyses contain several control variables. Gender is coded one for legislators who are male and zero for female. Party is coded one for legislators who are Republican, zero for Democrats. Legislative Branch is coded one for legislators who are Senators, zero for legislators who are House representatives. If the legislator held a leadership position at any time, he or she scores one on Leader, otherwise zero. The variable Years in Office is an interval level variable that indicates the total number of years that the legislator served between the first election to office and the end of the 52nd Legislature in 2011. Also, if the legislator served less than the full term limit of 12 years (applicable in Oklahoma to both houses of the legislature) before reaching the end of the period examined (again, the 52nd Legislature), he or she scores a one for the variable Early Exit. For example, if a legislator was elected in 2002 and left office at the end of the 51st Legislature in 2009, that legislature scores a one for "Early Exit." If he/she was elected in 2006 and was still in office at the end of the 52nd Legislature (2011), then he/she scores a zero. Likewise, if the legislator was elected in 1998 and served through 2010, he/she also scores a zero as he/she served the full 12-year term limit. If the legislator represents a district that is in the greater Oklahoma City or Tulsa metropolitan areas, that legislator scores a one for the variable Metro Area, otherwise zero.

Each legislator also has an interval score for two variables expressing proportionate demographics in his or her respective district. The Percent Native American in District indicates the percentage of that legislator's constituency (in the district he or she represents) that identified themselves as Native American in the 2010 Decennial Census. Because Oklahoma has a large Native American population, and tobacco issues in the state are closely tied to the concerns of Native American communities, we identify this group rather than other minorities. However, we also include the Percent Other Nonwhite in District, which measures the proportion of all nonwhites except Native Americans in an individual legislator's home district.

RESULTS

Table 6.1 presents bivariate Pearson correlations between all the independent and dependent variables. Significant correlations suggest the likelihood with which one variable correlates, or "explains" the variation, in another. In this subsection, we discuss the most statistically significant findings.

First, consider the strongest correlations—identified as being below the .001 level of significance. The greater the percentage Native Americans within a district, the less likely the district is to have a Republican legislator. Districts with greater percentages of Native Americans are also less likely to be located in the Oklahoma City or Tulsa metro areas. Greater proportions of other nonwhites in a district increases the likelihood of having a female legislator, while these districts are more likely to be in a metro area and metro areas are more likely to elect Republicans. A low score on the tobacco control policy score, interpreted as being more pro-tobacco, is more likely to be assigned to a Republican.

Several correlations that are not quite as strong as those above are still significant at the .01 level. The amount of tobacco campaign contributions received by a legislator increases the likelihood of his/her early exit. We theorize that some legislators loyal to tobacco companies leave the legislature in order to become industry lobbyists, as many of the lobbyists discussed in chapter 3 were former Senators or Representatives. Male legislators are more likely to receive higher amounts of tobacco contributions than female legislators, however only 20 out of 170 legislators are women, so any relationship between gender and another variable is susceptible to outliers. Metro areas are more likely to have legislators with a lower tobacco policy control score. Finally, higher amounts of tobacco campaign money are associated with lower tobacco policy scores.

The variables strongly affecting scores in table 6.1's bivariate relationships, Metro Area and Tobacco Contributions, both of which lowered scores, appear to be functions of political party. In the bivariate correlations, districts in the Oklahoma City and Tulsa areas are more likely to elect Republicans,

Table 6.1. Bivariate Pearson Correlations

		1	2	3	4	5	6	7	8	9	10	11
Gender (male)	1	1.00										
Party (Republican)	2	0.02	1.00									
Branch (Senate)	3	−0.03	−0.05	1.00								
Leader	4	−0.14†	0.01	0.02	1.00							
Years in Office	5	−0.03	−0.10	0.06	0.25***	1.00						
Early Exit	6	−0.05	−0.15*	−0.02	0.05	0.06	1.00					
Metro Area	7	−0.12	0.25***	0.09	−0.12	−0.05	0.06	1.00				
Percent Native American	8	0.15*	−0.42***	0.00	−0.01	−0.04	0.02	−0.39***	1.00			
Percent Other Nonwhite	9	−0.26***	−0.12	−0.01	−0.11	−0.02	−0.06	0.28***	−0.42***	1.00		
Tobacco Contributions	10	0.18**	0.15*	−0.05	0.13†	0.05	0.24**	0.07	−0.02	−0.10	1.00	
Tobacco Policy Score	11	−0.13†	−0.41***	−0.06	−0.02	−0.09	0.01	−0.20**	0.14†	0.03	−0.18**	1.00

* < .05, ** < .01, *** <.001, † <.10

Table 6.2. Stepwise OLS Regressions: Dependent Variable: Tobacco Policy Score. Unstandardized (B) and Standardized (STB) Coefficients

	Model 1		*Model 2*		*Model 3*		*Model 4*		*Model 5*	
	B	*STB*	*B*	*STB*	*B*	*STB*	*B*	*STB*	*B*	*STB*
Intercept	6.16***	0	7.37***	0	7.67***	0	8.42***	0	10.03***	0
Tobacco Contributions	–0.01*	–0.17	–0.00	–0.10	–0.00	–0.12	–0.00	–0.09	–0.00	–0.8
Gender (male)	–0.92	–0.10	–0.95	–0.11	–1.00	–0.11	–1.03	–0.12	–1.27*	–0.14
Party (Republican)			–	–0.40	–2.29***	–0.39	–2.40***	–0.41	–.58***	–0.45
Branch (Senate)					–0.58	–0.09	–0.54	–0.09	–0.49	–0.08
Leader					–0.11	–0.19	0.07	0.01	–0.09	–0.02
Years in Office							–0.12†	–0.13	–0.13†	–0.14
Early Exit							–0.25	–0.03	–0.27	–0.04
Metro Area									–0.70	–0.12
Percent Native American									–0.05	–0.11
Percent Other Nonwhite									–0.02	–0.09
N	170		170		170		170		170	
R^2	.04		.19		.20		.22		.24	

* < .05, ** < .01, *** <.001, † <.10

and legislators representing metro area districts tend to have lower tobacco control scores.

MULTIVARIATE ANALYSIS

Table 6.2 shows the results of ordinary least squares regression of policy scores regressed on campaign contributions and control variables.

When metro area and political party are included together in multivariate analyses, only party remains significant, indicating that legislators representing metro areas have lower scores only because they are more likely to be Republicans. A similar dynamic appears with regard to campaign contributions. Republicans have slightly higher average campaign amounts, and higher campaign amounts are modestly associated with lower scores. However, being a Republican is highly predictive of a lower score (this correlation is the strongest between any two variables in the study), and when both party and campaign contributions are included together the effect of contributions disappears.

CONCLUSION

Our most basic finding is that the link between campaign contributions and tobacco policy score is spurious. Receipt of more tobacco money only predicts a lower tobacco policy score because Republicans get more money and they score lower. After 2003 to 2010, health advocates ceased engaging in highly public advocacy to pressure the state legislature from the outside by holding politicians publicly accountable for their pro-tobacco actions. Instead, insider power politics became the primary focus of advocacy, which resulted in little gains for tobacco control in Oklahoma. The primary reason for this lack of progress was the party identity of the majority Republican Party in Oklahoma's legislature.

Chapter Seven

Lessons Learned

Public Ethics and the Public Good

The story of tobacco control in the state of Oklahoma during the early 2000s is that of a long overdue turnaround. Prior to 2001, the state of Oklahoma's record of health and tobacco use coincides with the dominant position of tobacco industry lobbying that had characterized the state's political climate for most of the period since the mid-1980s. Oklahoma's per capita cigarette sales were 27 percent higher than the national average in 2002, and over 26 percent of Oklahoman's identified themselves as cigarette smokers in 2004 (compared to the 23 percent national average). For each dollar spent on tobacco prevention and cessation in Oklahoma, the industry has spent about $35 for marketing to support policies favorable to tobacco use.

This dismal state of affairs was illustrated by the Oklahoma Alliance on Health or Tobacco's 1999 state tobacco policy plan, as well as the Oklahoma Tobacco Use Prevention and Cessation Advisory Committee's state plan in 2002. Both reports recommended a variety of policy proposals, including stiffer sanctions on youth access violations, repealing state preemption laws, adopting clean indoor air laws for public places and workplaces, and conducting vigorous anti-tobacco counter-marketing campaigns. Despite the ambitious recommendations of these plans, and the organizations' advocates having achieved some limited progress during the 1990s, tobacco industry lobbying and public relations campaigning dominated legislative activity and effectively prevented most tobacco control proposals from advancing through the legislature. Thus, real legislative gains in substantive tobacco control were almost non-existent prior to 2002. From 2003 onward, however, remarkable strides were made in a few policy areas such as clean indoor air, increased tobacco taxes, and youth access restrictions.

A number of specific lobbyists and allied organizations established a confident hegemony over the state law-making process with regard to tobacco issues, for which their industry clients continually praised them and their success. The major catalyst for strengthening the effectiveness of tobacco control advocacy was the arrival in 2001 of Dr. Leslie Beitsch, M.D., J.D. as the new commissioner of health. Determined to actively promote tobacco control policies as directed by the Oklahoma State Board of Health, and frustrated by the lack of legislative progress, Commissioner Beitsch promulgated, with assistance in one degree or another from organized health interests, the implementation of Board of Health administrative rules to establish clean-indoor air in public places and workplaces. While many of the rules went unsigned by the governor due to confliction with current state statute, the controversy incited by the rules' enactment, as well as Beitsch's brash and outspoken public relations style, precipitated a successful clean indoor air bill in the 2003 legislature.

For 17 years after Oklahoma raised the cigarette tax in 1987 from 18 to 23 cents, proposals to raise the tax were effectively squashed by industry lobby efforts. Finally, in 2004 the legislature placed a state question on the November 2004 election ballot for a 55 cent per pack net increase in the cigarette excise tax. The referendum passed, despite over $1.7 million spent by the industry-funded vote-no campaign. Before the measure took effect, Oklahoma ranked 42nd in the nation with regard to the amount of its cigarette excise tax. After the referendum, it moved up to 15th in the nation.

The fervor and publicity generated by Dr. Beitsch's aggressive outsider approach to enact more vigorous clean indoor air legislation, which was a catalyst for the passage of successful tobacco tax increase in 2004, helped move the struggle for tobacco control away from conventional insider lobbying and negotiation and into the public spotlight. Health organization and tobacco control advocates could advance successful efforts at rigorous legislation by augmenting their usual insider lobbying activity with a variety of aggressive public relations and outside political organization activities. These activities can include ballot initiatives, public rallies, letters to newspaper editors, paid issue ads, and involvement with possible community forums, and politicians who support industry interests can be revealed and held publicly accountable.

After Dr. Beitsch resigned in 2003, efforts toward clean air stalled. Although restaurants were required to build separately ventilated smoking rooms in 2006, smoking was banned at zoos in 2007, and ten college campuses were smoke-free by 2011, Oklahoma received a "D" grade in 2010 from the American Lung Association because of unrestricted tobacco use in bars, special restaurant rooms, and casinos; and restricted but not prohibited smoking on the grounds of private workplaces and government worksites. House Bill 2135, which would have reversed the preemption language of the 1987 Smoking In Public Places Act, was defeated during the 2010–2011 legislative sessions and identical legislation was defeated in 2012.

From 2005 to 2010, tobacco industry political action committees gave $75,875 to political campaigns, political parties, and political events. Of that total amount, $49,375 went to individual legislators' campaigns. Male legislators and Republican legislators tend to receive more campaign money from tobacco PACs, and a Republican party membership is significantly associated with being more pro-tobacco.

Despite the post-2003 slump in clean indoor air efforts in Oklahoma, public opinion continued to back more stringent anti-tobacco legislation, with 67 percent of Oklahomans surveyed in 2008 believing that no smoking should be allowed in restaurants at all (even in smoking rooms), 74.4 percent agreeing that no smoking should be allowed at all in shopping centers, and 69 percent believing that breathing someone else's cigarette smoke is very harmful to their own health (Stearns 2011).

PUBLIC ETHICS AND SERVING THE PUBLIC GOOD

In 1985, Dennis Thompson asked the provocative question: "Is administrative ethics possible?" He answered this important question by posing two ethical courses of action for public administrators, including those involved in tobacco control. The first course of action is when administrators are ethical by remaining loyal to the mandates of enabling legislation. While administrators can use discretion in determining how to implement a public law, they must always follow the mandates and requirements of those who enact public policies. Thompson criticizes this approach, arguing that administrators cannot always be neutral because they have competing obligations to colleagues, their agency, and society as a whole (Thompson 1985b). Additionally, the ethics of neutrality limit managers to obedience or resignation. Thompson rejects this either–or dichotomy because there are ways an administrator can dissent but continue to be employed by the agency.

A second course of public ethics action by administrators is based on moral action for the government or the society. In this case, administrators have no accountability for personal moral judgment and individual actions. Instead, their ethical responsibilities are a fixture of their job requirements for a public agency. Thompson also criticizes this perspective, arguing that personal responsibility is a requirement to maintain legal and democratic accountability.

To further this argument, Thompson advanced the notion that there are paradoxes in government ethics, including other major public policy issues of greater importance than government ethics and differences between personal and governmental ethics (Thompson 1985a). Gormley examined the role of individual motivation in public ethics by classifying public administrators as moralists who follow the law, pragmatists who balance obedience with self-interest, or rogues who lie and disobey orders (Gormley 2001). Maletz and Herbel found that ethics in government can only occur modestly through ethics requirements, education, and reform efforts. They concluded that democratic politics are a potent means to check self-interest and ensure ethics in government (Maletz and Herbel 2000).

The ASPA Code of Ethics codifies the notion that the democratic process is an important factor in ensuring public ethics, as in tobacco control (American Society of Public Administration 2012). More specifically, the code calls for interactive participation in response to citizen demands in the policymaking process, support for the public's right to know, making decisions that may not be popular, and compassion, benevolence, fairness, and optimism in interactions with citizens (Gormley 2001).

Only from 2001 through 2003 did Oklahoma made significant progress in legislative action to promote clean indoor air in public places and workplaces, and in significantly raising excise taxes. The dismal lack of progress and complete industry hegemony that characterized the legislative climate for the previous decade begs important questions. What changed? How did the industry so easily maintain its position over the years, and how did health interests suddenly begin attaining achievements from 2001 to 2003? Most importantly, what lessons does this history suggest about the future of health advocacy and tobacco policy making in Oklahoma and elsewhere?

During the late 1980s and throughout the 1990s, industry control of the legislative climate in Oklahoma was maintained by the skillful use of centralized lobbying, campaign contribution making, and support building among allied organizations like the hospitality industry. In their lobbying of state legislators, industry supremacy easily dwarfed the efforts of health organizations and advocates by maintaining an impenetrable insider relationship with key decision makers. That this level of access was combined with campaign contributions assured that lobbying efforts on behalf of tobacco control would be of little to no consequence. As Barbara Kumpe, advocacy director for the American Heart Association in Arkansas and Oklahoma, pointed out, "They do have deep pockets, I mean they have a lot of money that we don't have and as the Heart Association we do not give money to any candidates whatsoever . . . it's a policy, we do not fund anybody, and we do not give any money to their [legislators] campaign" (Kumpe 2004). Philip Morris and R. J. Reynolds do give money to politicians' campaigns, and they often hire the legislature's respected former colleagues to communicate the industry's interests. Such insider tactics were clearly the industry's strength and the health organizations' weakness. These insider strengths were furthermore buttressed by the industry's ability to mobilize outside pressure during their calling campaigns that identified industry-sympathetic voters and directed them to their representatives' offices.

The level of extreme disparity between the relative influences of the tobacco industry and health advocacy in Oklahoma was largely the result of the health organizations' approach to tobacco control advocacy, which was to conduct mostly conventional insider lobbying activities. Unfortunately, the proverbial playing field of conventional insider lobbying with regard to tobacco control legislation was completely dominated by the powerful and wealthy tobacco industry. Moreover, since preemption of clean air had been secured in the 1980s, all efforts had to be confined to the level of the state legislature, where the industry enjoyed its supremacy.

Dr. Leslie Beitsch's arrival in 2001, and his subsequent shifting of the playing field from legislative lobbying to high profile external pressure, demonstrated a stark contrast between the glacial pace of tobacco control progress that occurred prior to 2001 and the rapid achievements that he, in tandem with the health organizations to one degree or another, would make possible over the following two years. Beitsch's ability as Health Commissioner to promote public ethics and the public good through the Oklahoma Board of Health's administrative rules, thus forcing the issue of clean-indoor-air in the public realm and onto the legislature, was an important part of the strategy that culminated in SJR 21, but just as important was his vehement public campaigning on tobacco control issues such as a tobacco excise tax increase. By circumventing the usual quiet processes of lobbying committee chairs and other legislators, sitting in meetings with tobacco lobbyists and attempting to negotiate compromises, strategies that by themselves had produced few results in previous years, Beitsch moved tobacco control to the public arena, consistent with the public administration ethics of promoting the public good, and did so with a combative stance and with some disregard for alienation. His efforts paid off as the next three years saw legislation that would have seemed previously out of reach.

Barbara Kumpe described the contrast between Beitsch's approach and that maintained by health organizations prior to 2001, stating that, "Since the resignation of Dr. Leslie Beitsch, the smoke free movement has accomplished relatively little, in particular repealing local clean indoor preemption" (Kumpe 2004). Meanwhile, although Oklahoma tobacco consumption rates have lowered, the state continues to be well above national averages in proportions of adult (as well as most high school and middle school) smokers. For the goal of clean air, these conditions could be improved considerably through approaches like the repeal of state preemption of stronger local ordinances.

It is important to recognize, as well, the sociocultural context of Oklahoma history. Like many states across the trans-Mississippi West and the Far West of North America, the state was largely established by white and Native settlers intent on carving their own paths, valuing individualistic ethics by claiming, and taming, their own land (Madsen 2010). The pro-tobacco argument has historically been framed as a matter of personal liberty and individual choice, that is, Big Government attempting to control the smoker. This social construction of liberty is difficult to attack by focusing only on the effects of secondhand smoke. All of the factors mentioned above play an important role in the political success that the tobacco industry has seen in Oklahoma.

LESSONS LEARNED

Overall, from 1985 to 2001, the efforts of tobacco industry lobbying and public relations campaigning was clearly instrumental in countering the struggle by public health interests in tobacco control in a variety of issues, including excise taxes and clean indoor air laws. With the state health department's utilization (assisted by the health groups) of aggressive outsider tactics like public forums, issue advertisements, and free media interviews, in tandem with traditional insider lobbying approaches from 2001 to 2004, Oklahoma tobacco control in such areas as tobacco taxes and state clean indoor air advanced considerably. This change was due to the use of astute outsider tactics in conjunction with insider lobbying approaches. Prior to 2001, health advocates relied strictly on insider approaches in which they could never successfully outmatch the powerful tobacco industry on its own turf.

We do not suggest that health organizations abandon conventional insider lobbying activities, or that they discontinue meeting with and arranging legislation with legislators. Rather, they should combine such activities in tandem with a variety of aggressive public relations and outside political organizing activities to supplement traditional insider lobbying.

Initiatives can be used to propose and enact legislation that completely bypasses the legislature. Oklahoma statutory initiatives, for example, can be brought to the ballot with a signatory threshold of 8 percent of votes cast in the last state election (for the state office receiving the largest number of votes, generally the governor). The benefit of initiative campaigns is twofold. First, by running an initiative, there is almost always no need for compromise legislation from the state legislature, where industry power is most consolidated. Industry influence outside the legislature is uncertain, and in the past has centered on galvanizing behind-the-scenes supporters to call their legislators and using allied organizations for public announcements and lobbying that would look suspicious coming directly from the industry. The strenuous industry-funded efforts that were employed against SQ 713 in November 2004, amounting to almost $2 million in public campaigning against the bill, were unsuccessful in part due to strong public support for vigorous tobacco control policies and health programs.

Second, initiative campaigns may draw public and media attention with greater prominence than would be the case in legislative proposals, and as was the case with the publicity surrounding the health department's administrative rules, such public attention may compel officials to address the issue by enacting legislation in the following session. As the industry has maintained such a stronghold on preemption at the state level, initiatives may be an effective way to bring the preemption issue to Oklahoma voters, if health advocates can raise the exorbitant financial costs.

Health groups and other advocates should also significantly increase and employ as many astute external public relations and political tactics as possible, especially when legislative bills or initiative campaigns are in progress, including public rallies at the capitol, letters to newspaper editors, paid issue ads, and involvement with possible community forums. Through these methods, politicians who support industry interests should be held publicly accountable. We believe that these methods will dramatically enhance efforts to enact tobacco control legislation, as demonstrated between 2002 and 2004, when the public fervor generated by Dr. Beitsch's aggressive approach augmented health organizations' efforts and moved the struggle outside conventional insider negotiation and lobbying.

Through these insider and outsider public campaigning techniques, the public health community can continue the winning record established between 2002 and 2004 by strengthening current clean indoor air and tobacco taxes and enacting effective tobacco control legislation in a variety of other areas. This should include, as noted in the May 2002 *Oklahoma State Plan* efforts to: strengthen restrictions on tobacco sales in vending machines, stiffen penalties and enforcement for tobacco sales to minors, repeal all state laws that preempt stricter local tobacco control laws, adopt stronger clean indoor air laws in workplaces, establish effective statewide counter-marketing advertisements, create effective cessation programs, and establish vigorous anti-tobacco education campaigns.

Local preemption language should be reversed and repealed. This deficiency facilitates the ability of the tobacco industry to concentrate and consolidate all efforts at the state capital level. If preemption were reversed, lobbyists and PACs would be forced to deal with every city and small town in Oklahoma, perhaps lessening their influence and effectiveness.

Cigarette taxes were not raised between 2004 and 2010. Oklahoma's tax per pack is still below the national average, and should be raised to meet or exceed national averages to encourage a decrease in consumption.

One factor contributing to Oklahoma's grade of "D" from the American Lung Association is due to the state's permissiveness toward smoking in restaurants. Approximately 100–150 of these smoking rooms were built (totaling 1–2 percent of all Oklahoma restaurants) (Colberg 2011; Laux 2008), featuring separate ventilation and closure from main dining areas. However, research shows that toxin levels in these restaurants as a whole can be above a healthy level. Weak legislation passed in 2010 offers restaurants a 50 percent reimbursement (less depreciation) to dismantle their smoking rooms, but, so far, this law is making minuscule gains. Stronger legislation is needed to prohibit smoking from all indoor areas of restaurants. Public opinion supports this move.

Finally, legislators must be held politically and publicly responsible for accepting money and gifts from tobacco companies in return for ensuring unrestricted freedom in the state. Studies offer mixed conclusions on the existence of quid pro quo politics, yet maintaining a publicized list of "tobacco friendly" legislators might be a small step in reducing the variety of influences the tobacco industry wields at the capitol.

In conclusion, the story of Oklahoma's tobacco control politics is one that vividly illustrates the connection between public administration theory and actual real-life policymaking. The bureaucratic activism that Dr. Beitsch employed, particularly in using the emergency powers of the office of health commissioner to create laws (albeit temporary ones that served mainly to incite the public and put pressure on the legislature) in support of a public health initiative, was a stark departure from the passive role of public administrators evident in his Oklahoma predecessors, as well as his Florida contemporaries. The result in the early 2000s was a major battle victory in Oklahoma, first (and most significantly) with clean indoor air, and second with a long overdue cigarette tax increase. Ultimately, the benefit of the new outsider activist approach amounts to more than a set of policy gains in one state-level tobacco control struggle, although Oklahomans owe a great deal to Beitsch and the work he did. The heartland tobacco war has taught us that massively well-financed, exquisitely sophisticated, and seemingly invincible political insiders, who maintain an iron grip on policymaking for literally decades, *can* be overcome and defeated, but to do so we may need to broaden our understanding of the roles that public administrators, bureaucratic agencies, non-profit organizations, and other activists can, and should, play in the struggle to influence policy and lawmaking in the future.

References

American Cancer Society. 1998. *Analysis of Oklahoma Tobacco Documents*. Compiled and written by Prospect Associates.

American Lung Association. 2010a. *Colleges and Universities With 100% Tobacco Free Campuses*. Portland, Oregon.

———. 2010b. *State Laws Restricting Smoking by Location.*

American Society of Public Administration. 2012. *Code of Ethics.*

Associated Press. 1999. "Tobacco Increase Rejected," June 17.

———. 2001. "$1 Per Pack Cigarette Tax Proposed." December 12.

———. 2001a "Health Commissioner Bans Smoking at Departmental Buildings," *Associated Press State and Local Wire*. September 27.

———. 2002a. "Retail Outlets That Sell Cigarettes Oppose Proposed Cigarette Tax Increase."

———. 2002b. "Bill Would Ban Smoking in Public Places. *Associated Press State & Local Wire*, January 25.

———. 2002c. "Edmond Lawmaker Accuses House Leaders of Sabotaging Smoking Bill." February 12.

———. 2002d. "Man Files Lawsuit Over Smoking." *Associated Press State and Local Wire*, March 20.

———. 2004. "House Committee Narrowly Passes Cigarette Tax Increase," February 18.

———. 2007. "Oklahoma Store Owners Seek Changes in Tobacco Tax Stamp Rules." *Associated Press*, October.

———. 2006. "Penis Pump Judge Gets 4-Year Jail Term." *USA Today*, Retrieved from USA30.USAToday.com.

Balabanova, S. 1994. "Tabak in Europa Vor Kolumbus." *Antike Welt*. 75, pp. 282–286.

Balabanova, S., F. Parsche, and W. Pirsig. 1992. "First Identification of Drugs in Egyptian Mummies." *Naturwissenschaften* no. 79:358.

Barber, Brian. 2002. "Smoke-rules Foes Want Out of Federal Court." *Tulsa World*, July 2.

Barnes, Richard. 2001a. "The Shameful History of Secondhand Tobacco Smoke Control in Oklahoma. American Lung Association of Oklahoma, Oklahoma City, OK.

———. 2001b. "Use of Existing Powers of the Oklahoma State Board of Health and State Commissioner of Health to Regulate Smoking in Public Places," American Lung Association of Oklahoma, Oklahoma City, OK: February.

———. 2003a. "Action Alert on February 20 Committee Meeting," Email Communication. Oklahoma Alliance on Health or Tobacco. Oklahoma City, OK: February 21.

———. 2003b. "Conflicts and Problems with the Committee Substitute for SJR21. Email Communication. Oklahoma Alliance on Health or Tobacco. Oklahoma City, OK: February 21.

———. 2003c. "Letter to Governor Frank Keating. Oklahoma Alliance on Health or Tobacco." Oklahoma City, OK: March 18.

———. 2003d. "The Power of 'Public Nuisance' in the Smoking Restriction." Oklahoma Alliance on Health or Tobacco. Oklahoma City, OK: March 28.

———. 2004a. "Five Legislative Sessions That Changed Tobacco Control in Oklahoma."

———. 2004b. Telephone Interview with Andrew Spivak. September 23.

Beitsch, Leslie M. 2012a. Personal Interview with Andrew Spivak, Tallahassee, Florida: August 22.

———. 2012b. Telephone Interview with Michael Givel and Andrew Spivak, November 15.

Bellamy, Clayton. 2002a. "Judge Sends Smoking Rules Lawsuit Back to State Court," *Associated Press State & Local Wire*. July 12.

———. 2002b. "Judge Temporarily Blocks State's Rules Limiting Smoking in Restaurants," *Associated Press State & Local Wire*. July 16.

———. 2002c. "Restaurants Sue to Block New Smoking Rules," *Associated Press State & Local Wire*. June 27.

Bisbee, Julie. 2002. "Judge Denies Temporary Restraining Order on Smoking Rules," *Associated Press State & Local Wire*. July 1.

———.2003. "Tobacco Lobby Hits Bump in Road With Bills Passage Out of Committee," *Associated Press State & Local Wire*. February 21.

Bitton, A., Fichtenberg, C., and Glantz, S. 2001. "Reducing Smoking Prevalence to 10% in Five Years." *The Journal of the American Medical Association* no. 286 (21):2733–2734.

Borelli, Thomas. 2002. *Heads Up*. Philip Morris. Minneapolis: Minnesota Tobacco Depository. Bates No.: AC5000902664.

Borio, Gene. 2007. Tobacco Timeline, http://www.tobacco.org/resources/history/Tobacco_History.html/.

Burbridge, N. T. 1960. "The Australian Species of Nicotiana L. (Solanaceae)." *Australian Journal of Botany* no. 8:342–378.

Campaign for Tobacco Free Kids. 2010. Fact Sheet: Trends in Average State Cigarette Tax Rates. http://www.tobaccofreekids.org/research/factsheets/pdf/0211.pdf.

Carter, Ray. 2003a. "Legislators Urged to Find $41 Million for Trauma Care," *Journal Record Legislative Report*, News Section. April 6.

———. 2003b. "Senate Leader Seeks Support From Cigarette Tax Increase," *Journal Record Legislative Report*, News Section. April 16.

———. 2004a. "Democrats Reject Republican Proposal to Dramatically Raise Cigarette Tax," *Journal Record Legislative Report*, News Section. May 13.

———. 2004b. "Governor Henry Seeks Bipartisan Support for Cigarette Tax Proposal," *Journal Record Legislative Report*, News Section. February 6.

———. 2004c. "Health Care Groups Seek $1-per-pack Hike in Cigarette Taxes," *Journal Record Legislative Report*, News Section. January 22.

———. 2004d. "Officials Note Flaws in Cigarette Tax Proposal," *Journal Record Legislative Report*, News Section. April 27.

———. 2004e. "Oklahoma GOP Offers to Trade Cigarette Tax Hike for Capital Gains Cut," *Journal Record Legislative Report*, News Section. March 5.

———. 2004f. "Oklahoma House Backs Cigarette Tax Vote," *Journal Record Legislative Report*, News Section. May 21.

———. 2004g. "Oklahoma House Passes Shell Form of Cigarette Tax Bill," *Journal Record Legislative Report*, News Section. March 11.

———. 2004h. "Retailers Band Together to Battle Proposed Tobacco Tax Increase," *Journal Record Legislative Report*, News Section. September 8.

———. 2004i. "State Question 713 Supporters Say Tobacco Tax Increase Will Improve Oklahoma's Health Status," *Journal Record-News*, News Section.

Center for Public Integrity. 2006. Hired Guns. http://projects.publicintegrity.org/hiredguns/reg.aspx?state=OK

Chaloupka, Frank, Michael Cummings, Chris Morely, and Judith Horan. 2002. "Tax, Price and Cigarette Smoking: Evidence from the Tobacco Documents and Implications for Tobacco Company Marketing Strategies," *Tobacco Control* no. 11:62i–72i.

Clarkson, J. J., S. Knapp, V. Garcia, R. Olmstead, A. Leitsch, and M. Chase. 2004. "Phylogenetic Relationships in Nicotiana (Solanaceae) Inferred From Multiple Plastid DNA Regions," *Molecular Phylogenetics and Evolution* no. 33:75–90.

Colberg, Sonya. 2010. "New Law Could Further Limit Smoking Opportunities in Oklahoma City Restaurants," *The Oklahoman*, May 17.

———. 2011. "Oklahoma Board OKs Use of Tobacco Tax to Abolish Smoking Rooms," *The Oklahoman*, March 9.

Crutcher, Michael. 2001. *Tobacco Cessation Services*. Deputy Commissioner of Health and State Epidemiologist, October 11.

Dillow, Gordon. 1981. "The Hundred-Year War Against the Cigarette," *American Heritage Magazine*, February-March.

Edmondson, Drew. 2004. Personal Interview with Andrew Spivak. Oklahoma City, OK: July 27.

Ellis, Randy. 2002. "License Suspended for Role in Health Scandal," *Daily Oklahoman*: City Edition: March 15.

English, Paul. 2002. "Henry: Smoking Ban 'Makes No Sense'," *Tulsa World*: Saturday Final Home Edition, December 14.

———. 2003. "Cigarette Tax Hike Favored in Survey," *Tulsa World*, April 22, A9.

———. 2004. "Senate Passes Tobacco Tax Plan," *Tulsa World*, May 22, A1.

Estus, John. 2009. "Oklahoma Anti-Smoking Proposal Expected Next Session," *The Oklahoman*, December 26.

———. 2011. "Tobacco Lobbyists Step Up Work in Oklahoma Legislature," *The Oklahoman*, March 9.

Ervin, Chuck. 2002a. "Gutted Bill Passes Senate, But Without Much Steam," *Tulsa World*, March 21.

———. 2002b. "Keating Signs State No-smoking Bill," *Tulsa World*, April 18.

Ford, Brian. 2001. "State Curb on Smoking Ban Due Challenge." *Tulsa World*, September 14.

———. 2002a. "Officials Work to Widen Smoking Bans," *Tulsa World*, January 12.

———. 2002b. "Smoking Restrictions Get Put Out," *Tulsa World*, January 25.

———. 2002c. "Dust-Off: Committee Snuffs Smoking Bill," *Tulsa World*, January 31.

———. 2002d. "Anti-smoking Measure, Rules Add Fuel to Capitol Debates," *Tulsa World*, March 28.

———. 2002e. "Smoking Restrictions: Tobacco Lobby Has Clout," *Tulsa World*, April 7.

Francis-Smith, Janice. 2002. "Oklahoma Restaurant Association Wants Election on Smoking Ban," *Oklahoma Business News*.

———. 2004. "New Survey Finds Many Oklahomans Still Back Lottery," *Journal Record Legislative Report*, News section.

———. 2006. "Family Members Lobby Together at Oklahoma State Capitol," *Journal Record*, News section.

Givel, Michael S., and Andrew L. Spivak. 2008. "Public Management and the Public Good: The Case of Oklahoma's 2002 Secondhand Tobacco Smoke Rules," *Politics & Policy* 36(3): 420–447.

Givel, Michael S., and Andrew L. Spivak. Winter 2011–2012. "Bureaucratic Advocacy and Ethics: A State-Level Case of Public Agency Rulemaking and Tobacco Control Policy," *Public Integrity* no. 14 (1):5–18.

Glantz, Stanton, and Edith Balbach. 2000. *Tobacco War: Inside the California Battles*. Berkeley: University of California Press.

Goodspeed, Thomas. 1954. *The Genus Nicotiana; Origins, Relationships, and Evolution of Its Species in the Light of Their Distribution, Morphology, and Cytogenetics*. Waltham, Massachusetts: Chronica Botanica Co.

Gormley, W. T. 2001. "Moralists, Pragmatists, and Rogues: Bureaucrats in Modern Mysteries," *Public Administration Review* no. 61 (2):184–193.

Green, Richard. 2001. "New Partnership Announced to Fight Tobacco Addiction," *Associated Press State & Local Wire*.

Harper, David. 2002. "The Disabilities Act Does Not Preempt Smoking in Public Places, A Judge Says," *Tulsa World*.

Hertneky, Dana. 2012. "Oklahoma Lawmaker Kills Smoking Ban Bill," *Tulsa World*, April 17.

Hoberock, Barbara. 2002. "Anti-Smoking Bill May be Relighted," *Tulsa World*, March 23.

Hoberock, Barbara. 2010. "Smoke-Free Oklahoma Coalition Pushes Smoking Ban," *Tulsa World*, February 22.

Hoberock, Barbara and Paul English. 2003. "$1-A-Pack Cigarette Tax Hike Drive Begun," *Tulsa World*, April 16, A13.

Horton, P. 1981. "A Taxonomic Revision of Nicotiana (Solanaceae) in Australia," *Journal of Adelaide Botanical Gardens* no. 3 (1):1–56.

Humphreys, George. 2000. "A Century to Remember." Oklahoma City, Oklahoma: Oklahoma State Legislature.

Jenkins, Ron. 1999a. "Bond Program Could Climb to $500 Million," *Associated Press State and Local Wire,* State and Regional. April 22.

———. 1999b. "House Republicans Spoilers in Budget Battles," *Associated Press State and Local Wire*, State and Regional. April 26.

———. 2002a. "Ban on Smoking Extended in Bill," *Associated Press State & Local Wire*, February 14.

———. 2002b. "Supreme Court Rejects State Protest on Smoking Ruling," *Associated Press State & Local Wire*.

———. 2003. "New Health Chief Brings New Style to Post," *Associated Press State & Local Wire*. November 17.

———. 2004a. "Crucial Cigarette Tax Vote May Be Close," *Associated Press State and Local Wire*, State and Regional. September 12.

———. 2004b. "Oklahomans Like Cigarette Tax," *Associated Press State and Local Wire*, State and Regional. March 16.

Keating, Frank. 2002a. "Governor Takes Action on Restaurant and Other Public Place Smoking Rules." Office of Governor. Oklahoma City, OK, June 26.

———. 2002b. Letter from Governor Frank Keating to Oklahoma Commissioner of Health Leslie M. Beitsch, M.D., J.D.

Klein, Elizabeth, Jean Forster, Darin Erickson, Leslit Lytle, and Barbara Schillo. 2010. "Economic Effects of Clean Indoor Air Policies on Bar and Restaurant Employment in Minneapolis and St. Paul, Minnesota." *Journal of Public Health Management Practice* no. 16 (4):285–293.

Kumpe, Barbara. 2002. FW: press release on lawsuit. Email to Bob Miner. October 10.

———. 2004. Personal Interview with Andrew Spivak. Oklahoma City, OK: July 7.

Laux, Fritz. 2008. *Review of Economic Impact of Oklahoma's Smoke-free Air Policies on Restaurants.* Tahlequah, Oklahoma: Northeastern State University.

Lee, K., E. J. Hahn, H. E. Robertson, L. Whitten, L. K. Jones, and B. Zahn. 2010. "Air Quality In and Around Airport Enclosed Smoking Rooms." *Nicotine and Tobacco Research* no. 12 (6):665–668.

Leuthard, Joy. 2003a. URGENT!! LEGISALTIVE ALERT!!!!! ACTION NEEDED!!!. Email Communication. Oklahoma Alliance on Health or Tobacco. Oklahoma City, OK: February 27.

———. 2003b. GOOD NEWS!!!!!!!!!!! !!!. Email Communication. Oklahoma Alliance on Health or Tobacco. Oklahoma City, OK: March 10.

———. 2003c. TWO FACT SHEETS ATTACHED – URGENT 2 LEGISLATIVE ALERTS. Email Communication. Oklahoma Alliance on Health or Tobacco. Oklahoma City, OK: March 26.

———. 2003d. GOOD NEWS!! UPDATE ON TODAYS ACTIVITIES! AND LEGISLATIVE ALERT. Email Communication. Oklahoma Alliance on Health or Tobacco. Oklahoma City, OK: March 27.

———. 2003e. TWO LEGISALTIVE ALERTS!!!!!. Email Communication. Oklahoma Alliance on Health or Tobacco. Oklahoma City, OK: May 12.

———. 2003f. IMPORTANT: TUESDAY PRESS CONFERENCE ON CLEAN INDOOR AIR!. Email Communication. Oklahoma Alliance on Health or Tobacco. Oklahoma City, OK: May 19.

———. 2004. Interview With Andrew Spivak. Norman, OK.

Lorillard Tobacco. 1983. Exhibit B: Dear TAN volunteer. Minneapolis: Minnesota Tobacco Depository. Bates No.: 03669345

Lum, K. L., R. L. Barnes, and S. A. Glantz. 2009. "Enacting Tobacco Taxes by Direct Popular Vote in the United States: Lessons from 20 Years of Experience." *Tobacco Control* no. 18 (5):377–386.

Madsen, Deborah. 2010. "The West and Manifest Destiny." In *A Concise Companion to American Studies*, edited by John Rowe, 369–386. Hoboken, New Jersey: Wiley-Blackwell.

Maletz, D., and J. Herbel. 2000. "Beyond Idealism: Democracy and Ethics Reform." *The American Review of Public Administration* no. 30 (1):19–45.

Marks, Claire. 2007. "Evolution of Nicotiana L. (Solanaceae) in Australia." *Australian Systemic Botany Society Newsletter*, March, 4–6.

Matheny, Doug 2001. Re: Question regarding a news release. Email to Robert Miner, Janet Love, and Tracey Strader. Oklahoma State Department of Health. Oklahoma City, OK: January 16.

Matheny, Doug, and Robert Miner. 2004a. Personal Interview with Andrew Spivak. Oklahoma City, OK: July 15.

———. 2004b. Personal Interview with Andrew Spivak. Oklahoma City, OK: July 15.

McClave, Annette, Natalie Whitney, Stacy Thorne, Peter Mariolis, Shanta Dube, and Martha Engstrom. 2010. "Adult Tobacco Survey—19 States, 2003–2007, Surveillance Summaries." *Morbidity and Mortality Weekly Report* no. 59 (SSo3):1–74.

McGrew, Alice. 1996. "Narrative Application: SmokeLess Oklahoma Project," Grant proposal to the American Cancer Society, p.1.

———. 2004. Personal Interview with Andrew Spivak. Norman, OK: August 5.

Miner, Robert. 2001. P.O. for cessation medications. Email Communication to Julie Thompson, Julie Cox-Kain, Mike Crutcher, and Janet Love, Oklahoma State Department of Health: November 16.

———. 2002. Clean Indoor Air Communications Possibilities for October 10–11, 2002. Personal Notes. Oklahoma City, OK.

———. 2004. Personal Interview with Andrew Spivak. Oklahoma City, OK: July 15.

Moore, N. 1993. "Drugs in Ancient Populations." *Lancet*. 341:1157.

Morgan, David, Robert England, and George Humphreys. 1991. *Oklahoma Politics and Policies*. Lincoln, Nebraska: University of Nebraska Press.

Morris, Philip. (1997). Draft Oklahoma State Plan. Minneapolis: Minnesota Tobacco Depository. Bates No.: 2062900557, p.1.

Nichols, Suzanne. 2002. Rules. Oklahoma State Department of Health, Oklahoma City, OK: January 7.

Oklahoma Administrative Procedures Act. 2004. p. 106: Pursuant to Oklahoma Statutes, 250 (250.1 through 323).

Oklahoma Alliance on Health or Tobacco. 2004a. Oklahoma Alliance on Health or Tobacco: Partnership for a Healthier Oklahoma, Oklahoma City, OK.

———. 2004b. Publication of the Oklahoma Alliance on Health or Tobacco. State Ranking Document for Tobacco Taxes.

———. 2004c. State Ranking Document for Tobacco Taxes, Oklahoma Alliance on Health or Tobacco.

Oklahoma Attorney General. 2008a. *AG's Office Goes Tobacco Free*. Oklahoma City, OK.

———. 2008b. Oklahoma State Board of Health, http://www.ok.gov/health/Organization/Board_of_Health/.

Oklahoma Department of Libraries. 2007a. *Oklahoma Almanac*. Oklahoma City: Oklahoma Department of Libraries.

———. 2007b. *Oklahoma Almanac 2007–2008*. 51 ed. Oklahoma City: Oklahoma Department of Libraries.

Oklahoma Ethics Commission. 1999a. Form L-1 (January 5, 1999–January 31, 2003) Lobbyist Registration. Oklahoma City, Oklahoma. Oklahoma Ethics Commission.

———. 1999b. Form L-1 (January 12, 1999–January 6, 2003) Lobbyist Registration. Oklahoma City, Oklahoma. Oklahoma Ethics Commission.

———. 1999c. Form L-1 (January 19, 1999–January 24, 2004) Lobbyist Registration. Oklahoma City, Oklahoma. Oklahoma Ethics Commission.

———. 2002. Form L-2 (January 15, 2003–January 17, 2011).

———. 2002a. Form L-2 (January 17, 2001–January 15, 2003) Lobbyist or Other Person Gift Report, Attachment. Oklahoma City, Oklahoma. Oklahoma Ethics Commission.

———. 2002b. Form L-2 (July 18, 2002) Lobbyist or Other Person Gift Report, Attachment. Oklahoma City, Oklahoma. Oklahoma Ethics Commission.

———. 2004a. Form C1-R (November 10, 2004). Campaign Contributions and Expenditures Report, Citizens for a Healthy Oklahoma, Schedules A, E, & F. Oklahoma City, Oklahoma: Oklahoma Ethics Commission.

———. 2004b. Form C1-R (September 10, 2004). Campaign Contributions and Expenditures Report, Citizens for a Healthy Oklahoma, Schedules A, E, & F. Oklahoma City, Oklahoma: Oklahoma Ethics Commission.

———. 2004c. Form C-1R (November 10, 2004). Campaign Contributions and Expenditures Report, Vote No on Question 713 Committee, Schedule A. Oklahoma City, Oklahoma: Oklahoma Ethics Commission.

———. 2004d. Form C-1R (November 10, 2004). Campaign Contributions and Expenditures Report, Vote No on Question 713 Committee, Schedule E. Oklahoma City, Oklahoma: Oklahoma Ethics Commission.

———. 2004e. Form C-1R (October 11, 2004). Campaign Contributions and Expenditures Report, Vote No on Question 713 Committee, Schedule D. Oklahoma City, Oklahoma: Oklahoma Ethics Commission.

———. 2004f. Form C-1R. Campaign Contributions and Expenditures Report, Citizens for a Health Oklahoma, Schedules A, E, & F. Oklahoma Ethics Commission.

———. 2004g. Form L-3 (January 1, 2004) Termination of Lobbyist Registration. Oklahoma Ethics Commission.

———. 2004h. Form SO-2 (July 19, 2004). Statement of Organization For Committees, Citizens for a Healthy Oklahoma. Oklahoma City, Oklahoma: Oklahoma Ethics Commission.

———. 2004i. Form SO-2 (October 22, 2004). Statement of Organization for Committees, Vote No on State Question 713. Oklahoma City, Oklahoma: Oklahoma Ethics Commission.

———. 2010a. *2010 Ethics Manual for State Officers and State Employees*. Edited by State of Oklahoma. Oklahoma City.

———. 2010b. *Constitutional and Statutory Provisions and Constitutional Ethics Rules*. Oklahoma City.

———. 2010c. *Oklahoma Ethics Commission*. Edited by State of Oklahoma. Oklahoma City.

———. 2011. *Oklahoma Registered Lobbyists List/Registered Principal List*. Edited by State of Oklahoma. Oklahoma City.

Oklahoma Health Care Authority. 2010. Ready to Quit Smoking?, http://www.okhca.org/individuals.aspx?id=2733.

Oklahoma Insurance Department. 2007. Tobacco Settlement Endowment Trust Partnership. http://www.ok.gov/oid/Consumers/Public_Policy_Initiatives/Tobacco_Settlement_Endowment_Trust_Partnership/index.html.

Oklahoma Restaurant Association. 2002a. Letter to Dr. Ron Graves. Oklahoma City, OK: March 8.

———. 2002b. Oklahoma Restaurant Association Proposes Public Vote on Smoking. Oklahoma Restaurant Association.

Oklahoma Senate. 1987. Smoking in Public Places Act.

Oklahoma State Board of Health. 2002. Minutes. Oklahoma City, OK: June 6.

Oklahoma State Constitution, Limitation of Time Served in the Legislature.

Oklahoma State Department of Health. 2001. *Tobacco Free Policy*. Number 1–8, September.

Oklahoma State Department of Health. 2001. *Smoking in Public Places*. Oklahoma City.

———. 2001a. *Cessation Medications Regimens, Revised Cost Estimates*, November 12, 2001, Oklahoma City.

———. 2001b. *Congratulations*! Oklahoma City.

———. 2002. *Rule Impact Statement Secondhand Tobacco Smoke* 310:99 [NEW] Before the Oklahoma State Board of Health, June 6, 2002. Oklahoma City, OK.

———. 2004. Cigarette Sales Per Capita—2002. http://www.health.state.ok.us/program/tobac/factsand.htm

———. 2007. Key Facts. http://www.ok.gov/health/Disease,_Prevention,_Preparedness/Tobacco_Use_Prevention_Service/Facts_and_Data/Key_Facts.html.

———. 2009. Oklahoma Certified Healthy Business. Retrieved from http://www.ok.gov/health/Community_Health/Community_Development_Service/Turning_Point/Certified_Healthy_Business/ on June 18, 2010.

———. 2012. *Smoking Among Adults*. Oklahoma City.

Oklahoma State Legislature. 2010. Oklahoma State Legislature. http://newlsb.lsb.state.ok.us/.

Oklahoma State Senate. 2003a. Enrolled Joint Resolution 21, Oklahoma City, Oklahoma.

———. 2003b. Legislative Brief: Smoking in Public Places and Indoor Workplaces. July.

———. 2007. Enrolled Senate Bill 473. Oklahoma City, Oklahoma.

Oklahoma Statutes, Article XLII, Section 2532, 1915.

Oklahoma Tobacco Settlement Endowment Trust Fund. 2002. Fiscal Year 2002 Annual Report.

———. January 2008. Press Release. *Oklahoma Takes Next Big Step in Tobacco Control Program*. Oklahoma City, Oklahoma.

Oklahoma Tobacco Use Prevention and Cessation Advisory Committee. 2002. Oklahoma State Plan for Tobacco Use Prevention and Cessation.

Oklahoma Wholesale Marketers Association. 2010. About Us. http://www.owmanet.org/AboutUs/tabid/54/Default.aspx.

Outwater, Mary. 2004. *Tobacco Use and Legislation Survey*. University of Oklahoma Public Opinion Learning Laboratory (OU POLL).

Parsche, F., S. Balabanova, and W. Pirsig. 1993. "Drugs in Ancient Populations." *Lancet* no. 341 (8843):503.

Philip Morris. 1989. Summary of States' Political Status, Lobbyists. Retrieved and released by Doctors Ought to Care.

———. 1992. *Impact of Workplace Restriction on Consumption and Incidence*. Minneapolis, Minnesota: Minnesota Tobacco Document Depository; Bates No.: Bates No.: 2045447779.

———. 1996a. *Dear Customer*. Minneapolis: Minnesota Tobacco Depository. Bates No.: 2061855278.

———. 1996b. *Statement from the Oklahoma Restaurant Association*. Minnesota Tobacco Depository. Bates No.: 2061855159.

———. 1996c. *In Opposition to Efforts to Repeal Oklahoma's Uniform Smoking and Tobacco Sales Laws*. Bates No.: 2078842349.

———. 1996d. Oklahoma Association of Tobacco & Candy Distributers. Minneapolis: Minnesota Tobacco Depository. Bates No.: 2061855277.

———. 1998. *Oklahoma State Plan - 1999*. Minneapolis: Minnesota Tobacco Depository. Bates No.: 2074874356.

———. 1999a. *Oklahoma State Plan - 1999*. Minneapolis: Minnesota Tobacco Depository. Bates No.: 2074874356.

———. 1999b. *Script for Calling Voters*. Minnesota Tobacco Depository. Bates No.: 2080422444.

———. 2001. *Heads Up*. Edited by Thomas Borelli. Minneapolis: Minnesota Tobacco Depository. Bates No.: 5000905251-5000905262.

———. 2002. *Draft Phone Mobe* [sic] *Script*. Minneapolis: Minnesota Tobacco Depository. Bates No.: 2085757016.

Price, Marie. 2002a. "Legislators Summoned for Failure to Hear Smoking Ban Legislation." *Journal Record Legislative Report*, March 25.

———. 2002b "Group Submits Petition for Review of Oklahoma House Smoking Policy." *Journal Record Legislative Report*, March 25.

———. 2002c. "Subcommittee questions health commissioner on proposed smoking ban." Journal Record Legislative Report, March 28.

———. 2002d. "Oklahoma House Passes Smoking Ban for State Offices." *Journal Record Legislative Report*, April 10.

Rice, Misti. (2003). *American Cancer Society Letter to Dr. Beitsch*. Tulsa, OK: February 6.

R. J. Reynolds. 1985. Oklahoma: A State Analysis. R. J. Reynolds. Minnesota Tobacco Depository. Bates No.: 50509 1301-5.

———. 1988a. Letter to Mandel Matheson. R. J. Reynolds. Minnesota Tobacco Depository. Bates No.: 507633993.

———. 1988b. Choice Action Alert: Oklahoma. R. J. Reynolds. Minnesota Tobacco Depository. Bates No.: 507667765.

———. 1989. Oklahoma: July 1989, Smokers Made the Difference. Minnesota Tobacco Depository. Bates No.: 507669710.

———. 1990. State Tax Plan: Oklahoma. Minnesota Tobacco Depository. Bates No.: 507592151-507592152.

———. 1994. Internal Memo, Shirley From Vickie. R. J. Reynolds. Minnesota Tobacco Depository. Bates No.: 512539286.

———. 1995a. Email from David Powers to Tammy Payne. R. J. Reynolds. Minnesota Tobacco Depository. Bates No.: 513968038.

———. 1995b. Email from David Powers to Hyde, Mozingo, Muarray, Philips, Wold and Donahue. R. J. Reynolds. Minnesota Tobacco Depository. Bates No.: 528348993.

———. 1995c. Field Report: Oklahoma Legislature. R. J. Reynolds. Minnesota Tobacco Depository. Bates No.: 515323367-68.

———. 1998. Oklahoma Smoker Script. Minnesota Tobacco Depository. Bates No.: 527929334.

———. 2001a. Confidential (For Response Only) Oklahoma Considering Indoor Smoking Ban. Minnesota Tobacco Depository. Bates No.: 52522 2424.

———. 2001b. Profile of State Tobacco Laws (Through December 31, 2001). Minneapolis: Minnesota Tobacco Depository. Bates No.: 525141247-525141658.

Rajasevasakta, and A. K. Yegna Narayan Aiyer. 1966. *Field Crops of India*. 6 ed. Bangalore, India: The Bangalore Printing and Publishing Co., Ltd.

Richardson Ziebart Consulting LLC. 1998. Oklahoma Tobacco Regulation and Control.

Rogers, Everett, and Jeffrey Peterson. 2008. "Diffusion of Clean Indoor Air Ordinances in the Southwestern United States." *Health Education & Behavior* no. 35 (5):683–697.

Sieber, James E. 2002. "Phony, hypocritical." A Letter to the Editor. The Daily Oklahoman. June 16.

Slaymaker, Nick. 2002. RE: review deadline for March board mtg. Oklahoma State Department of Health. Oklahoma City, OK: January 29.

Social Security Death Index. 2010. Find your ancestors in the Social Security Death Index (SSDI)!

Sorrels, Dan. 2004. Personal Interview with Andrew Spivak. Norman, OK: July 7.

Spivak, Andrew L., and Michael Givel. 2005. *From Industry Dominance to Legislative Progress: The Political and Public Health Struggle of Tobacco Control in Oklahoma.* Norman, Oklahoma: University of Oklahoma.

State of Oklahoma, 1996. 2nd Session of the 45th Legislature. House Bill 2493.

Stearns, Ami. 2011. Oklahoma Text Search and Retrieval System. Oklahoma State Legislature, http://www.oklegislature.gov/tsrs_os_oc.aspx

Swedish Match. 2010. Swedish Match. http://www.swedishmatch.com/.

Swisher. 2010. Welcome to Swishertown: Your Sweet Times Start Here. http://www.swisher.com/web/main/;jsessionid=F5D1B9A94D5A89C685D9FB57D6C31EB0.tcn2.

Talley, Tim. 2002a. "Committee Votes to Create Designated Smoking Areas in Capitol," *Associated Press State & Local Wire*. April 1.

———. 2002b. "Judge Issues Injunction Against Smoking Rules," *Associated Press and Local Wire.* July 23.

———. 2002c. "Lawmakers, Governor Question Nonsmoking Rule," *Associated Press State & Local Wire*. March 25.

———. 2002d. "Resignation Fuels Capitol Smoking Debate," *Associated Press State & Local Wire*, March 21.

———. 2002e. "Committee Votes to Create Designated Smoking Areas in Capitol," *Associated Press State & Local Wire*. April 1.

———. 2003. "Lawmaker Proposes Tobacco Legislation," *Associated Press State and Local Wire*, State and Regional. January 22.

———. 2004a. "Attorney General Says Big Tobacco Blowing Smoke on Tobacco Tax," *Associated Press State and Local Wire*, State and Regional. October 20.

———. 2004b. "Coalition Proposes $1 Tobacco Tax Increase," *Associated Press State and Local Wire*, State and Regional. January 21.

The Daily Oklahoman. 2002a. "Above the Law Zealots Taking Over Policy Process." July 21.

———. 2002b. "Cooking Up Confusion: Beitsch Rules Blindside Small Business." July 25.

———. 2002c. "Dr. Beitsch, Heal Thyself Commissioner His Own Worst Enemy." April 21.

———. 2002d. "The Shadow Legislature," June 28.

———. 2002e. "Snuffing Out Choice; Board's Smoking Ban Beyond Reason." March 17.

The Oklahoman. 1905a. "Cigarette Law is Being Evaded Successfully by the American Tobacco Company," *The Oklahoman,* September 14.

———. 1905b. "Cigarette Paper Free," *The Oklahoman*, August 11.

———. 1906. "Cigarette Law: Judge Irwin Asks For Evidence With Which to Indict Offenders," *The Oklahoman*, September 14 .

———. 1907. "Campaign Against Coffin Nails On." *The Oklahoman*, June 5.

———. 1914. "Would Fine Minors Who Are Caught With 'Makins,'" *The Oklahoman*, April 26.

———. 2010. "Editorial: Latest Anti-Smoking Push Not Likely to Fail in Legislature," *The Oklahoman.*

Thompson, D.F. 1985a. "Paradoxes of Government Ethics." *Public Administration Review* no. 52 (3):254–259.

———. 1985b. "The Possibility of Administrative Ethics." *Public Administration Review* no. 45 (5):555–561.

Tobacco Institute. 1988a. Cigarette Excise Tax Factsheet: Oklahoma. Tobacco Institute. Minnesota Tobacco Depository. Bates No.: TIOK0031335-TIOC0031336.

———. 1988b. Memorandum to Stan Boman from Paula Duhaime. Tobacco Institute. Minnesota Tobacco Depository. Bates No.: TIOK0031338.

———. 1990a. Evaluations—1990 Recommendations: Region VIII Legislative Counsel. Tobacco Institute. Minnesota Tobacco Depository. Bates No.: TIOK0024316.

———. 1990b. Memorandum to George Minshew from Terry Frakes. Tobacco Institute. Minnesota Tobacco Depository. Bates No.: TIOK0030351.

———. 1990c. State Tax Plan. Minnesota Tobacco Depository. Bates No.: 507592151.

———. 1992a. Oklahoma Legislative Session Wrap-Up Report. Minnesota Tobacco Depository. Bates No.: TIOK 0028967.

———. 1992b. Check Request for Legislative Session in Oklahoma. Minnesota Tobacco Depository. Bates No.: TIOK 0028944.

———. 1994. Oklahoma: Ken Nance, 1993 Evaluation—1994 Recommendations. Minnesota Tobacco Depository. Bates No.: TI1740-1179-81.

———. 1995a. Letter Agreement to Ken Nance. Minneapolis: Minnesota Tobacco Depository. Bates No.: TI638-0503.

———. 1995b. Letter to OACTD Director. Minneapolis: Minnesota Tobacco Depository. Bates No.: TI1638-0545.

Tulsa World. 1996a. "Anti-Smoking Bill Offered," *Tulsa World*, January 12. News: p. A–21

———. 1996b. "Tobacco Bill Amendment Called 'Out,'" *Tulsa World*, April 18, News: p. C8.

———. 2011. "Editorial: Gearing Up: Local Control Effort Advances," *Tulsa World*, January 13.

United States District Court for the Northern District of Oklahoma, 2002. Memorandum of defendants veterans of foreign wars, Hoyak, Ltd, and the Oklahoma Restaurant Association in opposition to plaintiffs' motion for summary declaratory judgment (2002., Case No. 02-CV-569 H (C).

US Tobacco Weekly. 1998. OK: Senate Passes Bill on Smoking Ordinances. Lorillard Tobacco. Minneapolis: Minnesota Tobacco Depository. Bates No.: 99229522.

———. 2002. OK; Restaurants Announces Health Groups' Support of Smoking Ban. Edited by Philip Morris. Minneapolis: Minnesota Tobacco Depository. Bates No.: AC5000905078.

Wedel, Steve. 2003. "Oklahoma Health Care Authority Seeks $1-Per-Pack Cigarette Tax Increase," *Journal Record Legislative Report*, News Section. April 11.

Wheeler, H. M. 1935. "Studies in Nicotiana II. A Taxonomic Survey of the Australasian Species." University of California Publications in Botany no. 18 (4):45–68.

Williams, Pamela. 2002. FW: Tobacco/ORA News Release. Email Communication. Oklahoma State Department of Health. Oklahoma City, OK: October 9.

Wilson's Oklahoma Statutes, Chapter 25, Article 58, Section 2656, 1901. Prohibiting Sale or Giving Away Cigarettes, etc., Misdemeanor; Penalty.

Index

ABC: Oklahoma State Agencies, Boards, and Commissions, 86
Adair, Larry, 34, 37, 38–39, 63, 64
Administrative Procedures Act, 40, 43
Albrecht, Roxie, 66
Alcoholic Beverage Laws Enforcement Commission (ABLE), 83
Alexander, Chad, 81
Altria Corporate Services, Inc., 24, 68
American Cancer Society: and cigarette tax legislation, 66; in Florida, 32–33; and *HB 2493*, 16; member of Oklahoma Alliance on Health or Tobacco, 11; support for secondhand smoke legislation, 53
American Heart Association: and cigarette tax legislation, 58; in Florida, 32–33; member of Oklahoma Alliance on Health or Tobacco, 11; negotiation with tobacco industry, 12–13; and Oklahoma Restaurant Association, 52–53; on tobacco industry campaign contributions, 96; 2002 goals report, 2–3
American Lung Association: "D" grade for Oklahoma (2010), 94, 99–100; in Florida, 32–33; and Leslie Beitsch, 36; member of Oklahoma Alliance on Health or Tobacco, 11; and preemption of local laws, 18; and Smoking in Public Places Act (1987), 14, 30, 31, 47; suit against Oklahoma Restaurant Association (2002), 6; support for secondhand smoke legislation, 29, 53; on tobacco industry control in Oklahoma, 12; 2002 report, 2–3
American Tobacco Company, 11
Americans with Disabilities Act: as basis for American Lung Association lawsuit (2002), 6, 47; lawsuit over emergency clean air rules, 46; lawsuit over *HB 2401*, 37
ASPA Code of Ethics, 96

Balbanova, S., 10
Barker, Jim, 57
Barnes, Richard: on anti-tobacco advocacy from 1990s–2001, 12, 31; on Commissioner of Health Beitsch, 36; and preemption, 18; support of increased cigarette taxes, 60; and "Unclean Indoor Air Act of 1987," 14, 30; and use of emergency rules on tobacco use, 40, 42, 44, 52, 53
Beauchemin, Ed, 24

Beitsch, Leslie: arrival in Oklahoma, 31–32, 94, 97; ban on smoking in health department offices, 34–35; ban on smoking in the state capitol, 39; career in Florida, 32–33; criticism over clean indoor air rules, 45; and increased cigarette taxes, 59–68; and lawsuit over emergency clean air rules, 47; letter from Govervor Keating on secondhand smoke rules, 43; public outsider campaign, 51–52; relations with legislature, 50; resignation of, 50; and *SJR 21*, 27, 48–49; and use of bureaucracy to influence policy, 2, 40–44, 100
Bellmon, Henry, 14, 29, 57
Benson, Lloyd, 18, 59
Blackburn, Debbie, 21
Brannin, Dan, 29
Brickner, T. J., 61
Broadway, Charles, 45
Broken Arrow Medical Center Foundation, 69
Brown & Williamson Tobacco Corp., 24, 35, 47, 60
bureaucratic activism, 2, 40–44, 100
Bush, Jeb, 33

California, tobacco control in, 3–5
capital gains tax, 64
capitol, state, smoking ban in, 36–39
Cattlemen's Steakhouse, 74
Certified Healthy Business Program (2003), 73
Chilcote, Samuel, 24
Cigar Association of America, 68, 83
cigarette taxes. *See* taxes, cigarette
cigarettes: Oklahoma per capita sales, 3; percentage of smokers in Oklahoma, 3–5; taxes on, 55–71
clean air legislation, attitudes toward, 75–78
Coble, Tom, 61
Coburn, Tom, 21
Coleman, Carolyn, 38–39
Commerce, Industry and Labor Committee, 22, 37, 38, 39
contributions, political, defined, 80
Conwood Tobacco Company LP, 68, 82, 83
costs of health care for tobacco-related illness, 5
Crain, Brian, 75
Crutcher, Mike, 62

deaths, smoking-related, 4–5
Dobelbower, Judy, 38
Donaldson, Eric, 81
Drogin, Jamie, 63
Duhaime, Paula, 56
Dunlap, Jim, 67

Edmondson, Drew, 6, 38, 46, 55
Edwards, David Stephen, 81
emergency rules on tobacco use, 40–44, 98–99
Equal Employment Opportunity Commission, 38
ethics, public, 95–100
excise tax. *See* taxes, cigarette

Fallin, Mary, 64
Fields, Lloyd, 37–38, 39
Freddie's Barbecue and Steakhouse, 45–47

gifts, political, defined, 80
Giles, Elizabeth, 81
Graves, Bill, 39, 65
Greater Oklahoma City Chamber of Commerce, 70
Gregory, Jay, 34
GRPA Agency, Ltd., 83

Haselwood, Richard, 81
Health Care Policy and Research (OSMA), 40–41
health care costs for smoking related illnesses, 5, 60, 66, 73

Health Department. *See* Oklahoma State Department of Health
health organizations, role of in clean air legislation, 51–53
Henry, Brad, 49, 62, 63, 64, 74
Herbert, Dave, 27
Hiett, Todd, 63
Hobson, Cal, 61–63, 66, 67
Holmes, Sven Erik, 47
Hopper, Jim, 60, 64
House Bill 1007, 12
House Bill 1325, 56
House Bill 1812, 15
House Bill 2129, 12, 30
House Bill 2135, 74–75, 94
House Bill 2267, 75
House Bill 2401, 37
House Bill 2493, 15, 24
House Bill 2660, 63, 66, 73
House Bill 2774, 74, 77
House Business and Commerce Committee, 15
House Joint Resolution 1010, 61
Huddleston, Richard, 20, 21, 25, 81–82

industry, tobacco: twentieth century system of control in Oklahoma, 11–13; dominance in the 1990s, 30–36; lobbyists (2001–2003), 21–28; political action committees, 85–91
initiatives, statutory, use of, 40–44, 98–99

John Middleton Co., 81, 83, 84
Johnson, Glen, 18, 21, 22
Jones, Courtney Lee, 82
Journal Legislative Report, 63

Keating, Frank, 6; and adoption of emergency rules for protection from secondhand smoke, 41–43, 44–47, 51; and cigarette tax legislation, 59, 62; and *HB 2493*, 16; and *SB 1553*, 39; stance on tobacco issues, 32; tobacco industry contributions to, 25
Krantwashl, Larry L., 37
Kumpe, Barbara, 12–13, 52, 96, 97

The Lancet, 10
Lane, Joe, 68
legislation: clean air legislation, 31, 75–78, 87. *See also* specific legislation
Leist, M. D., 37
Leuthard, Joy, 40–41, 62, 69
Lewis, Steve, 57
lobbyists, tobacco: Alexander, Chad, 81; allied organizations, 25; campaign contributions, 25; defined, 80; Donaldson, Eric, 81; Edwards, David Stephen, 81; gift report, 24; Giles, Elizabeth, 81; Haselwood, Richard, 81; Huddleston, Richard, 81–82; industry lobbyists from 2001–2003, 20–28; Jones, Courtney Lee, 82; and legislative contributions, 26; Milner, James, 82; Nance, Brian, 82; Nance, Ken, 82–83; Philip Morris, 19–20; Phillips, Richard, 83; Rasmussen, Dianne, 83; Rogers, Tom, 83; Stem, Robert, 84; twentieth-century system of control in Oklahoma, 11–13; Vanatta, Benny, 84; Webb, Nathan Thomas, 84; Williams, Don, 84–85
Lorillard Tobacco Company, 35, 37, 47, 55–56, 60, 68

Matheny, Doug, 18, 36, 40, 43, 48, 59, 62
Matheson, Mandell, 20, 20–21
Mazzio's Corporation, 27, 53, 74
McGrew, Alice, 11–12
Medicaid, 60, 66, 73
Milner, James, 82
Miner, Bob, 13, 18, 36, 41, 43, 48, 58
Mitchell, Bill, 65
Morgan, Fred, 59
Morgan, Mike, 46, 47

Morris, Ronald, 20, 25
Murcer, Bobby, 16

Nance, Brian, 37, 82
Nance, Ken: and ban of smoking in health department offices, 35; career as tobacco industry lobbyist, 22–24, 82–83; and cigarette tax legislation, 56, 57, 58, 60; and *House Bill 1812*, 15; and *State Question 713*, 68; and Tobacco Institute, 20
Native Americans and tobacco regulation, 13, 60–65, 74, 84
Nida, Jerry, 15, 36

Occupational Safety and Health Administration, 38
Oklahoma: political structure of, 35; sociocultural context of, 97–98
Oklahoma Adult Tobacco survey, 75–79
Oklahoma Alcoholic Beverage Law Enforcement Commission, 16
Oklahoma Alliance on Health or Tobacco (OAHT): and cigarette tax legislation, 55, 60, 62; *House Bill 1007*, 12; 1999 state tobacco policy plan, 93; organization of, 11; prohibition of tobacco use in day care centers (1992), 12; and *State Question 713*, 69; and use of emergency rules on tobacco use, 40
Oklahoma Association of Health Care Providers, 61, 69–70
Oklahoma Association of Tobacco and Candy Distributors, 20, 21, 24, 25, 58
Oklahoma Board of Health, 32, 35, 41
Oklahoma Cancer Society, 20
Oklahoma Department of Health: ban of smoking in health department offices, 34–35; corruption in, 33–34; 2002 goals report, 2–3
Oklahoma Department of Labor, 23, 38
Oklahoma Department of Libraries, 86
Oklahoma Ethics Commission, 25, 69, 79, 80, 86
Oklahoma Grocers Association, 16, 25, 60, 64
Oklahoma Health Care Authority, 61
Oklahoma Health Commission, 5–6
Oklahoma Health Commissioner: Beitsch, Leslie, 1, 31–54; Crutcher, Mike, 62; Nida, Jerry, 15, 36; and *State Question 713*, 70; under Oklahoma Board of Health, 35
Oklahoma Hospital Association, 69
Oklahoma Legislators' Tobacco Attitudes Scoring Sheet, 87
Oklahoma Lung Association, 20
Oklahoma Medicaid Program, 60, 66, 73
Oklahoma Oncology Inc., 69
Oklahoma Public Opinion Learning Laboratory, 4
Oklahoma Restaurant Association (ORA), 20, 25; and American Lung Association lawsuit, 47–48; Benny Vanatta, 24, 36, 84; and *HB 2493*, 16, 17; lawsuit against Oklahoma State Department of Health, 6, 45–47, 52; and Native American-owned casinos, 74; and secondhand tobacco smoke rules, 42, 44; and *SJR 21*, 6, 48–49; and state preemption of local laws, 27; tobacco industry use of, 31
Oklahoma Retail Grocers Association, 20
Oklahoma Retailers Alliance, 25
Oklahoma Smokers' Rights, 58–59
Oklahoma State Department of Health: and *House Bill 2129*, 30; lawsuit over emergency clean air rules, 45–47; member of Oklahoma Alliance on Health or Tobacco, 11; *State of the State's Health* (2004), 3
Oklahoma State Medical Association: and cigarette tax legislation, 62;

Health Care Policy and Research, 40–41; and Oklahoma Alliance on Health or Tobacco, 11; and State Question 713, 69; support for secondhand smoke legislation, 53
Oklahoma State Plan (2002), 99
Oklahoma Supreme Court, 47
Oklahoma Tobacco Settlement Endowment Trust, 6
Oklahoma Tobacco Use Prevention and Cessation Act, 34
Oklahoma Tobacco Use Prevention and Cessation Advisory Committee, 2–3, 93
Oklahoma Wholesale Marketers Association (OWMA), 85
The Oklahoman, 11, 45, 74

Payne, James H., 46
Payne, Tommy, 21
Petroleum Marketers Association, 60, 68
Philip Morris: Altria Corporate Services, Inc., 24, 68; campaign contributions, 96; and cigarette tax legislation, 59, 63; concern over workplace smoking restrictions (1992), 14; contributions to Governor Keating, 25; Draft Oklahoma State Plan (1997), 18; and *HB 2493*, 16–17; lobbyists, 19–20, 24, 81, 83, 84; and *State Question 713*, 68
Philips, Richard, 38, 65, 83
political action committees, 85–91
preemption, state, of local clean air laws, 2, 13–18, 99; and *HB 2493*, 17–18; and *SJR 21*, 27; and Smoking in Public Places Act (1987), 30, 74–75
Prevention of Youth Access to Tobacco Act, 16–17
Public Administration Ethics Code, 2
public administration theory, 1
public ethics, 95–100
Public Relations International, 68

QuikTrip, 60, 64, 66, 68

Rasmussen, Dianne, 83
restaurants: Cattlemen's Steakhouse, 74; and deadline to comply with *SJR 21*, 5; Freddie's Barbecue and Steakhouse, 45–47; and *House Bill 2129*, 12, 30; Mazzio's Corporation, 27; public attitudes toward smoking in, 76; and second hand tobacco smoke ordinances, 41–42; state's permissiveness toward smoking in, 99–100. *See also* Smoking in Public Places Act (1987)
Revenue and Taxation Committee, 64
Reynolds American, 81, 82
Reynolds Tobacco Company. *See* R. J. Reynolds Tobacco Company
Rice, Misti, 66
R. J. Reynolds Tobacco Company: and campaign contributions, 96; and cigarette tax legislation, 56–57, 58–59; contributions to Governor Keating, 25; and lawsuit over emergency clean air rules, 47; lobbyists, 21, 82; political analysis of Oklahoma (1985), 2; report on smoking regulation at local level, 15; and *SJR 21*, 27; and *State Question 713*, 68; State Tax Plan, 57–58
Roach, Russ, 38, 39
The Robert Wood Johnson Foundation, 58
Robinson, Ben, 25, 27, 38, 61
Rogers, Tom, 20, 24, 83

Schlim, Tex, 46
second hand tobacco smoke, 14, 29, 31, 38, 40–44, 50–51
Selby, Ken, 74
Senate Bill 1553, 38, 39, 44, 51
Senate Bill 327. *See* Smoking in Public Places Act (1987)
Senate Health Committee, 38

Senate Human Service Committee, 38
Senate Joint Resolution 21 (*SJR 21*), 5, 6, 27, 48, 50–51, 73, 74, 75, 78, 97
Sims, Dean, 68
Smokeless Oklahoma Program, 13
Smokeless Tobacco Council, 24
smoking, health impacts of, 4–5, 60, 66, 73
Smoking in Public Places Act (1987), 16–17; amended by *HB 2129*, 12, 30; changes in before passage, 14; enforcement and penalties for violation, 31; passage of, 29–30; and preemption of local laws, 74–75, 94; and protection from secondhand smoke, 41, 43; and Tobacco Institute, 15
Sorrels, Dan, 60
State Bill 473, 73
state capital gains tax, 64
State Question 640, 61
State Question 713, 55, 68–71, 73, 98
Steele, Kris, 74–75
Stem, Robert, 84
Stillwater Medical Center, 69
Stubbs, Dick, 74
Students Working Against Tobacco (SWAT), 13
Swedish Match, 82–83
Swisher International, Inc., 74, 84

Task Force on Tobacco and Youth, 59
taxes, cigarette, 55–71; *House Bill 1325*, 56; *House Bill 2660*, 63, 66; Oklahoma tax policy after 2004, 99; Oklahoma tax policy before 2001, 55–59, 94; *State Question 640*, 61; *State Question 713*, 68–71, 73; tobacco tax advocacy (2001–2004), 59–68
Thompson, Donald, 46
Thornbrugh, Mike, 60, 62, 64
Thurber, Mary, 20, 24, 25
Tobacco Action Network, 55–56
Tobacco Free Oklahoma Coalition (TFOC), 11, 27, 52, 58
tobacco, historical background of in Oklahoma, 9–19; anti-tobacco advocacy (1900–2001), 11–13; dominance in the 1990s, 30–36; in territorial era, 10–11
tobacco industry lobbying: from 2001–2003, 20–28; 20th century system of control in Oklahoma, 11–13; allied organizations, 25; campaign contributions, 25; effect on Oklahoma legislature, 2005–2010, 79–85; and emergency rules covering secondhand smoke, 44; gift report, 24; and legislative contributions, 26. *See also* lobbyists, tobacco
tobacco industry political action committees, 85–91
Tobacco Institute: commitment to preemption of smoking ordinances, 2, 15; lobbyists, 20, 22–23; "Regional Excise Tax Fact Sheet," 56; and *State Question 713*, 68; State Tax Plan, 57
Tobacco Merchants Association, 25
tobacco plant, 9–10
Tobacco Retailers Alliance, 68, 84
Tobacco Settlement Endowment Trust, 73
"Tobacco Stops With Me" campaign, 73
tobacco tax. *See* taxes, cigarette
Tobacco Use Prevention Service, 36
Tribal Retailers of Oklahoma, 84
tribal tobacco retailers, 60–66, 84
Tulsa World, 16, 37, 49, 62, 74
Tyler, Michael, 37

"Unclean Indoor Air Act of 1987," 14, 30
United States Tobacco (UST), 68
University of Oklahoma, 4, 12, 66, 69, 73
U.S. Smokeless Tobacco, 21, 81–82, 83, 84
U.S. Tobacco, 21

Vanatta, Benny, 24–25, 36, 44, 46, 47, 49, 74, 84
Vaughn, Raymond L., Jr., 17–18, 37, 38, 59–60, 61, 65
Veterans of Foreign Wars Post 1320, 45–47
Vincent, Bob, 40

Walgreen Company, 70
Walter, David, 23
Warden, Larry, 38
Webb, Nathan Thomas, 84
WGRC Inc., 84
William K. Warren Foundation, 69
Williams, Don, 84–85
Williamson, James, 66
Wilson Research Strategies, 65
Wright, John, 65

Xcaliber International, 81, 82, 84

Youth Access to Tobacco Act (1994), 71
youth, restriction on tobacco sales to, 15–16

About the Authors

Michael S. Givel is professor of political science at The University of Oklahoma-Norman. Professor Givel's teaching and research interests include: policy change theory, complexity theory and public policy, comparative public policy, public health policy, and tobacco policy. Professor Givel's research and peer-reviewed publications specifically focus on the evaluation of state functions, markets, corporate behavior and public policy and public administration at the local, state, federal, and international levels. This research has included examining the theoretical and applied nature of policy change and corporate or political actions particularly as it impacts tobacco policy. In 2009, Professor Givel was the first U.S. research and teaching Fulbright to Bhutan. While in Bhutan, he taught graduate classes in public administration and researched Bhutan's unique anti-tobacco law.

Andrew L. Spivak is assistant professor in the Department of Sociology at the University of Nevada, Las Vegas, where he teaches criminology, penology, demography, research methods, and statistics. A graduate of The University of Oklahoma (Ph.D., 2007), he previously worked in corrections for eight years. He is the author of *Sexual Violence: Beyond the Feminist-Evolutionary Debate* (2011, LFB Scholarly Publishing), and peer-reviewed articles published or forthcoming in *Social Science Quarterly, Crime & Delinquency, Deviant Behavior, Justice Research and Policy, Suicide and Life-Threatening Behavior, Urban Geography, Problems of Post-Communism*, the *Journal of Drug Education, Politics & Policy, Public Integrity*, and the *International Quarterly of Community Health Education.* His current research includes topics in prison recidivism, violent offending, juvenile justice processing, tobacco regulation, prostitution, and residential segregation.